84 Structure and Bonding

Springer

Berlin
Heidelberg
New York
Barcelona
Budapest
Hong Kong
London
Milan
Paris
Tokyo

Metal Complexes with Tetrapyrrole Ligands III

Volume Editor: J. W. Buchler

With contributions by
J. W. Buchler, C. Dreher, F. M. Künzel,
S. Licoccia, R. Paolesse, J. Šima

With 50 Figures and 36 Tables

Springer

Volume Editor:
Prof. Dr. J. W. Buchler
Technische Hochschule Darmstadt
Institut für Anorganische Chemie
Eduard-Zintl-Institut
Hochschulstraße 10, D-64289 Darmstadt

ISBN 3-540-59281-4 Springer-Verlag Berlin Heidelberg New York

Die Deutsche Bibliothek - CIP-Einheitsaufnahme
Metal complexes with tetrapyrrole ligands / vol. ed.:J. W. Buchler.
Berlin; Heidelberg; New York; Barcelona; Budapest; Hong Kong; London; Milan; Paris; Tokyo: Springer.
Literaturangaben
NE: Buchler, Johann W. (Hrsg.)
3. With contributions by J. W. Buchler ... - 1995
(Structure and Bonding; 84)
ISBN 3-540-59281-4
NE: GT

Typesetting: Macmillan India Ltd., Bangalore-25, India
SPIN: 10477225 51/3020 - 5 4 3 2 1 0 Printed on acid-free paper

Editorial Board

Attention all "Structure and Bonding" readers:

A file with the complete volume indexes Vols. 1 through 84 in delimited ASCII format is available for downloading at no charge from the Springer EARN mailbox. Delimited ASCII format can be imported into most databanks.

The file has been compressed using the popular shareware program "PKZIP" (Trademark of PK ware Inc., PKZIP is a available from most BBS and shareware distributors).

This file is distributed without any expressed or implied warranty.

To receive this file send an e-mail message to:

SVSERV@DHDSPRI6.BITNET.

The message must be: "GET /CHEMISTRY/SB_V1.ZIP".

SVSERV is an automatic data distribution system. It responds to your message. The following commands are available:

HELP	returns a detailed instruction set for the use of SVSERV,
DIR (*name*)	returns a list of files available in the directory "name",
INDEX (*name*)	same as "DIR",
CD <*name*>	changes to directory "name",
SEND <*filename*>	invokes a message with the file "filename",
GET <*filename*>	same as "SEND".

Preface

The present Volume 84 of Structure and Bonding is entitled "Metal Complexes with Tetrapyrrole Ligands III" and completes a series of three volumes dedicated to this general topic which started with Volume 64 and continued with Volume 74. The first volume contained topics such as stereochemistry of metallotetrapyrroles, infrared and Raman spectra, biomimetic porphyrins, or metalloporphyrins with metal-carbon single bonds and metal-metal bonds. In the second volume, subjects like extended X-ray absorption fine structure or metal tetrapyrroles with special electrical and optical properties were covered.

This concluding volume contains three articles:

1. "Photochemistry of Tetrapyrole Complexes"
   (Jozef Šima)
2. "Metal Complexes of Corroles and Other Corrinoids"
   (Sylvia Licoccia and Roberto Paolesse)
3. Synthesis and Coordination Chemistry of Noble Metal Porphyrins"
   (Johann W. Buchler, Christine Dreher, and Frank M. Künzel)

The first article gives a broad review of photochemical reactions of metalloporphyrin and metallophthalocyanine systems. After a short introduction to the photophysics of the excited states of porphyrins and metolloporphyrins, a systematic presentation of the hitherto observed photochemical reactions of metallotetrapyrroles is given. The material is divided into individual chapters concerning photosubstitutions, photoeliminations, and photoadditions, photoinsertions, photoisomerizations, and photoredox reactions (the latter are reported separately for central atom or tertapyrrole centered processes). Photochemical formation and degradation of polynuclear complexes and applications of photoreactions in the tetrapyrrole field conclude the article. In order to restrict the size of the volume, the material is presented in 7 tables which give a quick overview and references. Jozef Šima is Assistant Professor of Inorganic Chemistry at the Slowak Technical University (Bratislava) and author of a recent textbook devoted to the photochemistry of coordination compounds (Ref. 1 of his article) appearing both in English and Slovakian.

The second article deals with the synthesis and properties of metal corrolates and corrinoids. In view of the many articles written on corrinoids, the corrolates are in the foreground of the article, a detailed discussion of their structures, of their IR-, UV/Vis-, photoelectron-, and NMR spectra, and of their electrochemical behavior being given. Sylvia Licoccia is Professor of Chemistry at the University of Rome Tor Vergata and has not only made notable scientific contributions to corrole chemistry, but to porphyrin chemistry as well.

The third article is devoted to the synthesis and chemical reactions of noble metal porphyrins. Especially the porphyrin derivatives of ruthenium, osmium, and rhodium have turned out to have a rich oxidation/reduction, hydride or organometallic chemistry which appeares to be very promising for the developement of catalytic industrial processes, either for hydrocarbon or olefin oxidation, or selective hydrocarbon synthesis. Furthermore, their chemistry serves to explain many reactions that are typical for heme proteins and vitamin B_{12}. The editor has taken the opportunity to write this article, together with two of his former doctoral students, and has tried to condense the reactions into schemes in order to allow easy comparison of the chemical reactions or porphyrins belonging to the Ru/Os or Rh/Ir dyads. He hopes that these schemes together with the explanatory tables are not too confusing to the readers, and that has not forgotten to mention any authors for their contributions to this field which is a rapidly developing research area.
The articles written for the previous volumes (Struct. Bonding Vols 64 and 74, having appeared 1987 and 1990, respectively) on "Metal Complexes with Tetrapyrrole Ligands" are also listed in the following Table of Contents.

Darmstadt, June 8, 1995 Johann W. Buchler

Table of Contents

Table of Contents of Volume 64

Table of Contents of Volume 74

Synthesis and Coordination Chemistry of Noble Metal Porphyrins

J. W. Buchler*, C. Dreher, F. M. Künzel

Institut für Anorganische Chemie, Technische Hochschule Darmstadt, Petersenstraße 18, D-64287 Darmstadt, Germany

The noble metals ruthenium, osmium, rhodium, iridium, palladium, platinum, silver, and gold form a variety of complexes M(P) LL′ with tetrapyrrole ligands which are studied in many laboratories because their coordination chemistry and catalytic power is similar to that of the corresponding iron and cobalt tetrapyrroles forming the prosthetic groups in heme proteins and vitamin B_{12}. Apart from the corrins and corroles which are treated in the second article of this volume, the main interest is devoted to noble metal complexes of porphyrins, but complexes of hydroporphyrins, porphyrin analogs, homologs, porphyrinoids, and phthalocyanines are also mentioned where appropriate.

The treatise is organized into four main topics: metal insertion ("equatorial coordination chemistry", Sect. 2, formation of the M(P) entity), substitution and redox reactions ("axial coordination chemistry", Sect. 3, exchange of axial ligands L and L′), organometallic chemistry (a special case of axial coordination chemistry, Sect. 4), and notes on current special chemical topics like electrochemistry, catalysis, supramolecular chemistry, and biological applications.

Where possible, the material is compressed in schemes and tables giving a synopsis of consecutive reactions of Ru/Os and Rh/Ir dyads. The two other dyads, Pd/Pt and Ag/Au, are so strongly governed by the square planar d^8 configuration that they are practically devoid of axial coordination chemistry. The redox reactions of the Ru/Os dyad live on the switches between d^6–d^5 and d^6–d^4–d^2 configurations mostly with octahedral geometry about the metal, those of the Rh/Ir dyad on the switches between the d^6–d^7–d^8 configurations, giving rise to octahedral, square-pyramidal, and square planar geometries. The occurrence of the dinuclear or mononuclear Rh d^7 systems give a special touch of radical chemistry to the organometallic chemistry of rhodium porphyrins which is expected to be fruitful to organometallic chemistry as a whole.

Although a broad overview was intended, many entries are just given in keywords or tables in order to give the reader a chance to get more information. Nevertheless, out of nearly 500 papers scrutinized, only nearly 400 references appear in order to keep the volume and the writing time within certain limits, and any research group missing a paper in this review is asked for patience with the authors.

*This article is dedicated to my friends and colleagues, Professors Dr. Peter Paetzold (Aachen) and Dr. Nils Wiberg (München) on the occasion of their sixtieth birthdays.

Structure and Bonding, Vol. 84

1 Introduction

1.1 Noble Metals As a Special Class of Transition Metals

"Noble metals", or "precious metals", in the elemental state are not attacked by atmospheric oxygen at ambient temperature and form metal oxides that easily disintegrate on heating [1, 2]. In the broadest sense [1], rhenium, the six platinum metals (ruthenium, osmium, rhodium, iridium, palladium, platinum – the first four of which are sometimes termed the rarer platinum metals [3]), silver, gold, and mercury are assembled in this class (Fig. 1) [1]. This assembly can be justified by the positive reduction potentials of its members, but less so by the lability of the corresponding oxides, Re_2O_7 boiling at 327 °C, ReO_3 subliming at 614 °C, and OsO_4 being formed from metallic osmium at 800 °C. (Lower osmium oxides and osmium containing compounds may react even at room temperature with atmospheric dioxygen to form volatile OsO_4, a fact that should be kept in mind when working with osmium compounds. OsO_4 vapours are known to cause severe eye damage.)

In a narrower sense, therefore, the class of noble metals encompasses the platinum metals as well as silver and gold in German usage (solid frame in Fig. 1) [4]. The Pauling electronegativities [5] of these metals range from 1.9 (Ag) to 2.4 (Au) and express a high tendency to form covalent bonds which is not expected by the lower Allred-Rochow electronegativities. English usage of the expression "noble metals" is less frequent [6].

Cr	Mn	Fe	Co	Ni	Cu	Zn	Ga
1.7	1.6	1.8	1.9	1.9	1.9	1.6	1.6
Mo	Tc	Ru	Rh	Pd	Ag	Cd	In
1.8	1.9	2.2	2.2	2.2	1.9	1.7	1.7
W	Re	Os	Ir	Pt	Au	Hg	Tl
1.7	1.8	2.2	2.2	2.2	2.4	1.8	1.7

Fig. 1. The periodic table of noble metals (in the *solid frame*) and their neighbouring elements including Pauling electronegativities. The so-called pseudo-noble metals appear in the *dashed frames*

This review is devoted to those porphyrinato metal complexes MLL′(P) (**1**) (for abbreviations see Table 1) that contain the abovementioned noble metals, M = Ru, Os, Rh, Ir, Pd, Pt, Ag, and Au.

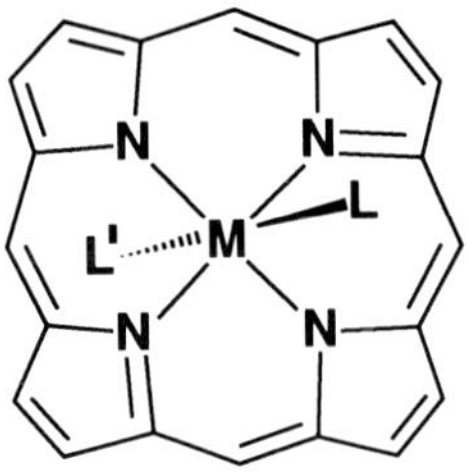

1: MLL′(P)

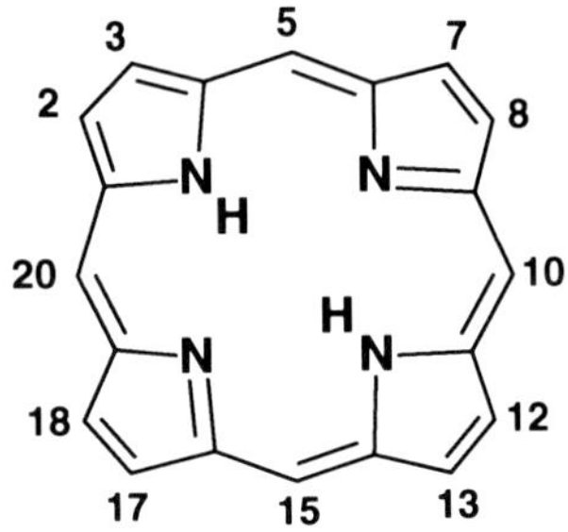

2: $H_2(P)$

1.2 Nomenclature, Abbreviations, and Coordination Chemistry of Metal Porphyrinates

Generally, metal porphyrinates (or frequently, but not conforming to nomenclature rules, "metalloporphyrins") contain a metal ion, M^{n+}, a porphyrinate ligand $(P)^{2-}$, and donor ligands L^{k-} which complete the coordination sphere, as required by the charge or the electronic configuration of the central metal ion [7, 8]. The porphyrin ligands $(P)^{2-}$ are introduced as their conjugate acids, the porphyrins $H_2(P)$ (**2**) which are specified by the letter codes given in Table 1. Some of these capital letter codes (OEP, TPP, TMP, TTP) are so widely used that they are noted in a recently published register of abbreviations [9]. Strict nomenclature [10], however, would require that the porphyrinates like any abbreviation of a ligand be quoted in minuscules, but $(tpp)^{2-}$ is only rarely seen in the literature; nevertheless, putting the letters in parentheses [10] and counting the negative charges is obligatory.

The coordination chemistry of porphyrins can be subdivided in equatorial and axial coordination chemistry as written in Eqs. (1) and (2) and depicted with the coordination types given in Fig. 3, i.e. square-planar **A**, square-pyramidal **B**, or distorted-octahedral **C**; the latter is illustrated by formula **1** [7, 8].

Axial coordination (X, univalent anion):

$$MX_2 + H_2(P) \rightleftarrows M(P) + 2HX \qquad (1)$$

Equatorial coordination (types indicated):

$$\underset{\mathbf{A}}{M(P)} \overset{L}{\rightleftarrows} \underset{\mathbf{B}}{ML(P)} \overset{L'}{\rightleftarrows} \underset{\mathbf{C}}{MLL'(P)} \qquad (2)$$

These reactions may be accompanied by redox reactions. The letter code ML(P)L′ of the octahedral coordination type **C**, or formula **1**, L written

Table 1. Abbreviations used in this article

a. Definition of the letter codes of the more common metal free porphyrins $H_2(P)$ (2) (porphyrin free acids)

Synthetic Porphyrins	
$H_2(P)$	Porphine (or any unspecified porphyrin)
$H_2(Pc)$	Phthalocyanine (5,10,15,20-tetraaza-2,3,7,8,12,13,17,18-tetrabenzporphyrin)
$H_2(Pyc)$	Porphycene
$H_2(DPB)$	1,8-Bis(2,8,13,17-tetraethyl-3,7,12,18-tetramethylporphyrin-5-yl)biphenylene
$H_2(Etio\ I)$	Etioporphyrin I
$H_2(a\text{-}H_4OEP)$	2,3,7,8,12,13,17,18-Octaethyl-2*H*,3*H*,7*H*,8*H*-porphyrin (octaethylisobacteriochlorin)
$H_2(H_2OEP)$	2,3,7,8,12,13,17,18-Octaethyl-2*H*,3*H*-porphyrin (octaethylchlorin)
$H_3(HOTPP)$	5-Hydroxy-5,10,15,20-tetraphenyl-5*H*,22*H*-porphyrin
$H_2(npOEP)$	5,15-Di(1-naphthyl)-2,3,7,8,12,13,17,18-octaethylporphyrin
H(NPhOEP)	21-Phenyl-2,3,7,8,12,13,17,18-octaethylporphyrin
$H_2(OEP)$	2,3,7,8,12,13,17,18-Octaethylporphyrin
$H_2(OEPAr)$	5-Aryl-2,3,7,8,12,13,17,18-octaethylporphyrin
$H_3(OEPH)$	2,3,7,8,12,13,17,18-octaethyl-5*H*,21*H*-porphyrin (Phlorin)
$H_2(OEPMe_2)$	5,15-Dimethyl-2,3,7,8,12,13,17,18-octaethyl-5*H*,15*H*-porphyrin
$H_2(OETAP)$	2,3,7,8,12,13,17,18-Octaethyl-5,10,15,20-tetraazaporphyrin
H(SDPDTP)	5,20-Diphenyl-10,15-ditolyl-21-thiaporphyrin
$H_2(TAP)$	5,10,15,20-Tetrakis(*p*-anisyl)porphyrin
$[H_2(TEAP)]^{4+}$	5,10,15,20-Tetrakis(4-{ethyldimethylammonio}phenyl)porphyrin
$H_2(TBP)$	Tetrabenzporphyrin
$H_2(TDBOHPP)$	5,10,15,20-Tetrakis(3,5-*tert*-butyl-4-hydroxyphenyl)porphyrin
$H_2(TH_4CPP)$	5,10,15,20-Tetrakis(4-carboxyphenyl)porphyrin
$H_2(TMeCPP)$	5,10,15,20-Tetrakis(4-methoxycarbonylphenyl)porphyrin
$H_2(TMP)$	5,10,15,20-Tetrakis(2,4,6-trimethylphenyl)porphyrin
$H_2(TMeP)$	5,10,15,20-Tetramethylporphyrin
$[H_2(TMPyP)]^{4+}$	5,10,15,20-Tetrakis(4-methylpyridinio)porphyrin
$H_2(ToAPP)$	5,10,15,20-Tetrakis(*o*-aminophenyl)porphyrin
$H_2(TpCF_3PP)$	5,10,15,20-Tetrakis(4-trifluormethylphenyl)porphyrin
$H_2(TPrPyc)$	2,7,12,17-Tetra-*n*-propylporphycene
$H_2(TPP)$	5,10,15,20-Tetraphenylporphyrin
$[H_2(TTMAP)]^{4+}$	5,10,15,20-Tetrakis(4-trimethylammoniophenyl)porphyrin
$H_2(TTP)$	5,10,15,20-Tetrakis(*p*-tolyl)porphyrin
$H_2(iPrTPP)$	5,10,15,20-Tetrakis(4-isopropylphenyl)porphyrin
$[H_2(TPPS_4)]^{4-}$	5,10,15,20-Tetrakis(4-sulfonatophenyl)porphyrin
$H_2(TTiPP)$	5,10,15,20-Tetrakis(2,4,6-triisopropylphenyl)porphyrin
$H_2(TXP)$	5,10,15,20-Tetrakis(3,5-dimethylphenyl)porphyrin
$H_2(\{MeO\}_{12}TPP)$	5,10,15,20-Tetrakis(2,4,6-trimethoxyphenyl)porphyrin
$H_2(3,4,5\text{-}MeOTPP)$	5,10,15,20-Tetrakis(3,4,5-trimethoxyphenyl)porphyrin
$H_2(TTEPP)$	5,10,15,20-Tetrakis(2,4,6-triethylphenyl)porphyrin
$H_2(TCl_8PP)$	5,10,15,20-Tetrakis(2,6-dichlorophenyl)porphyrin
Native Porphyrins	
$H_2(Deut\text{-}DME)$	Deuteroporphyrin IX-dimethylester
$H_2(Hemato\text{-}DEE)$	Hematoporphyrin IX-diethylester
$H_2(Meso)$	Mesoporphyrin IX
$H_2[(Meso)G]$	Apoglobin-mesoporphyrin IX complex
$H_2(Meso\text{-}DME)$	Mesoporphyrin IX-dimethylester
$H_2(Meso\text{-}DtBuE)$	Mesoporphyrin IX-di-*tert*-butylester
$H_2(Proto)$	Protoporphyrin IX
$H_2(Proto\text{-}DME)$	Protoporphyrin IX-dimethylester
$H_2(Uro\text{-}I)$	Uroporphyrin I

Table 1. Continued

b Metalloporphyrins

This part of the list just gives examples to describe the coordination type (**A**, **B**, **C**, . . .) with short formulae.

M(P)(**A**)	Metalloporphyrin with unspecified metal and without axial ligands
Pd(TPP)(**A**)	Tetraphenylporphyrinatopalladium(II)
RhMe(OEP)(**B**)	Methyl(octaethylporphyrinato)rhodium(III) (M = Rh, L = Me, P = OEP)
$PtCl_2$(TPP)(**C**)	Dichloro(tetra-*p*-tolylporphyrinato)platinum(IV) (M = Pt, L,L′ = Cl, P = TTP)
Os(1-Meim)$_2$(OEP)	Bis(1-methylimidazole) (octaethylporphyrinato)osmium(II) (M = Os, P = OEP, L, L′ = 1-Meim
RuCO(TPP)Py(C)	Carbonyl(tetraphenylporphyrinato)pyridineruthenium(II) (M = Ru, L = CO, P = TPP, L′ = Py)*
[OsOMe(OEP)]$_2$O (**CC**)	μ-Oxo-bis[methoxo(octaethylporphyrinato)osmium(IV)] (**CC** means two connected octahedral systems)

* *Note*: The more firmly bound axial ligand will appear before, the more labile axial ligand behind the porphyrin symbol which is always in parentheses. This avoids extra parentheses. Valency superscripts or valency notations in parentheses should not be given in formulae!

c. Anionic ligands X^-

Anion X^-	Name
$(mnt)^-$	Maleodinitrile dithiolate
Np^-	Naphthalenide
OAc^-	Acetate
OEt^-	Ethoxide
OMe^-	Methoxide
OPh^-	Phenoxide
OR^-	Alkoxide
R^-	general aliphatic anion
SR^-	Alkyl thiolate

d. Neutral donor ligands L, solvents, or substitutents

Abbreviation	Name
Bipy	4,4′Bipyridine
Bu	*n*-Butyl
*i*Bu, *t*Bu	*iso*-Butyl, *tert*-Butyl
1-*t*Bu-5-PhIm	1-*tert*-Butyl-5-phenylimidazole
4-*t*Bupy	4-*tert*-Butylpyridine
COD	Cyclooctadiene
COE	Cyclooctene
Cp	Cyclopentadienyl
Cp*	Pentamethylcyclopentadienyl
DEGE	Diethylene glycol monomethylether
Dib	Di(isocyano)benzene
DMSO	Dimethyl sulfoxide
2,3-DMI	2,3-Dimethylindole
dpm	Bis(diphenylphosphino)methane
dppe	1,2-Bis(diphenylphosphino)ethane
en	Ethylenediamine
Et	Ethyl

Table 1. Continued

EtOH	Ethanol
HOAc	Glacial acetic acid
HIm	Imidazole
Me	Methyl
Me_2NH	Dimethylamine
MeOH	Methanol
1-MeIm	1-Methylimidazole
2-MeTHF	2-Methyltetrahydrofuran
Ph	Phenyl
Pr	Propyl
Py	Pyridine
Pyz	1,4-Pyrazine
PPh_3	Triphenylphosphine
PR_3	Trialkylphosphine
$P(OR)_3$	Trialkylphosphite
RX	Alkyl halide
RNC	Alkyl isocyanide
THF	Tetrahydrofuran
THT	Tetrahydrothiophene

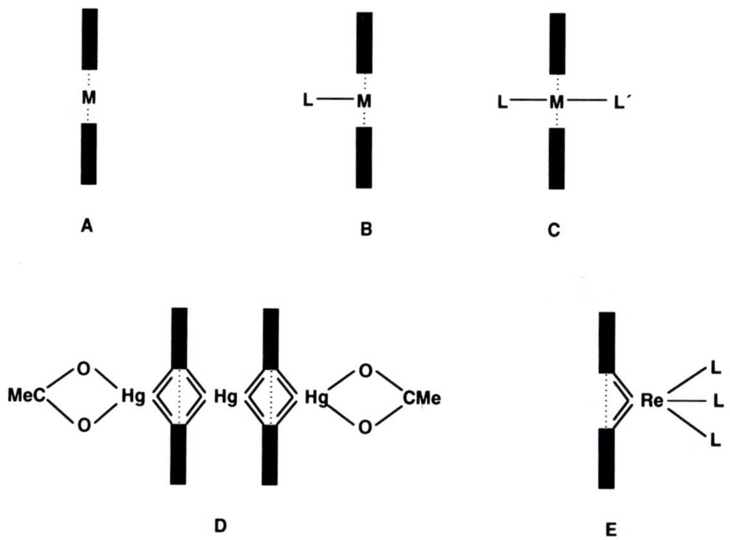

Fig. 2. Coordination types **A–E** of porphyrin complexes

immediately after M, means that a certain axial ligand L is more firmly bound to M than another one, L′; this abbreviation procedure requires the least amount of parentheses. If both ligands are equal, **C** is written as $ML_2(P)$.

1.3 Noble Metals As Defined by Their Behavior Inside a Porphyrinato Ligand

The statement that the coordination chemistry of the noble metal porphyrins is generally restricted to the coordination types **A** to **C** defines the borderline to the two "pseudo-noble metals" Re and Hg (broken frames in Fig. 1). In certain mercury porphyrins octacoordinate Hg(II) central ions occur, e.g. in the trinuclear complex AcOHg (OEP)Hg(OEP)HgOAc [11], type **D**. Recently, heptacoordinate Re(V) porphyrins, e.g. fac-$ReCl_3$ (TAP), type **E**, have been made [12]. Thus, the porphyrin ligand serves to discriminate the two "pseudo-noble metals" Re and Hg from the "genuine" eight noble metals Ru...Au using the statement that the latter do not normally occur with coordination numbers higher than 6.

Another common feature of noble metal porphyrins is the existence of derivatives with M(II) and M(III) central metal ions which show at the same time a very high stability towards mineral acids, i.e., the central metal ions are not removed by treatment of the complex with cold concentrated sulfuric acid, attempting a reverse of Eq. (1). Due to this behavior, these complexes belong to "stability class I" [7, 8]. It is not known whether the recently prepared bis(phosphane) Re(II) porphyrins [13] (low-spin d^5-configuration) also belong to stability class I, but structurally and chemically they are very similar to the corresponding Os(II) complexes (low-spin d^6-configuration). Hg(II) porphyrins (stability class II) are demetallated with HCl, anyway; they are thus far less stable towards acids than the porphyrin complexes of the divalent noble metals in the narrow sense. Indeed, the typical noble metal porphyrins have central metal ions with d^6 to d^9 electronic configurations which are prone to metal d_π to ligand π^* back donation, a circumstance that stabilizes the coplanar situation of the metal in the porphyrin ring or at least within the four central nitrogen atoms. This back-donation may be the reason why the complexes of noble metal porphyrins can only be demetalated under destruction of the macrocycle.

1.4 Reasons for Investigations on Noble Metal Porphyrins

The very important role of the heme system, Fe(Proto)LL′ (**3**) in biological oxygen transport and consumption as well as electron transport is a main topic not only of biochemists, but of bioinorganic chemists and biomimetic chemists as well; for this general topic, the reader may consult some recent review articles [14–21, 22]. Bioinorganic chemists have studied the effect of replacement of iron by other 3*d* metals, especially chromium, manganese, and cobalt, and frequently, interesting structural, spectral, or functional models [14, 20] of the heme enzymes have been found with these metals.

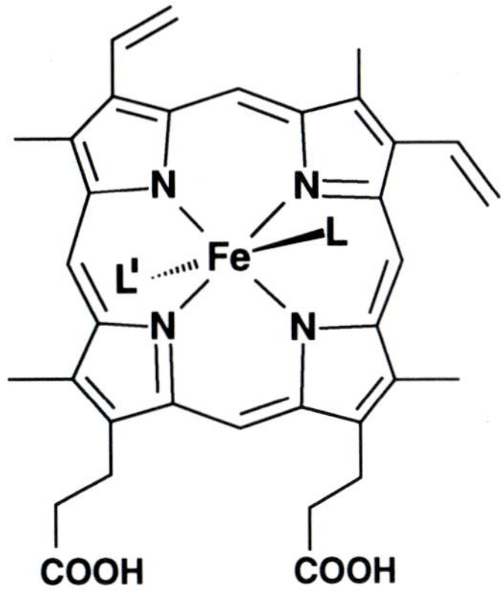

3: FeLL′(proto)

Replacement of the later 3*d* metals in the porphyrin systems, e.g. **1** or **3**, by their homologous 4*d* or 5*d* counterparts has the following consequences.

1. The covalent bonds to axial ligands L and L′ in **1** or **3** become stronger, hence the coordination sphere is more stable, and unstable intermediates or active species of reactions catalyzed by the heme systems may persist with the heavier homologs and easier lend themselves to chemical and physical investigations.
2. Due to the larger spatial extension of the 4*d* and 5*d* orbitals, metal d_π to ligand π^* bonding or its reverse are much more pronounced and its effects can be easier observed with the heavier transition metal porphyrins.
3. Due to the stronger covalent bonding to the central metal, the 4*d* or 5*d* metal porphyrins are much more stable against photochemical destruction or irreversible oxidation. Hence, dilute solutions are much more easily subject to photophysical investigations than the 3*d* counterparts. This is important for studying photochemical, photophysical, photoelectrical, or photocatalytical [23] properties which are becoming gradually more important with metalloporphyrins which are not only interesting as biomimetic systems, but likewise as novel electronic and optoelectronic materials [24–28].

For this reason, since the last general treatise on porphyrin complexes of the individual metals [8], many papers have appeared, and it seems worthwhile to discuss the chemistry of noble metal porphyrins separately from any general review on metal porphyrins on the basis of material preferentially published within the last fifteen years.

2 Synthesis of Noble Metal Tetrapyrroles (Metal Insertion)

This chapter deals with the provision of suitable starting materials for investigations in the noble metal porphyrin field (Sects. 2.1–2.4) and the establishment of their structures (Sect. 2.5). Most of these investigations are devoted to complexes with synthetic porphyrin ligands which are soluble in organic solvents [e.g. (OEP)- or (TPP) complexes; Sect. 2.1]. Special mention is made to some novel porphycene derivatives (Sect. 2.2), to phthalocyanine systems (Sect. 2.3), and to water-soluble porphyrin complexes (Sect. 2.4).

2.1 Survey of the Insertion of Individual Noble Metals

A decade ago, almost every metal had been inserted into a porphyrin, i.e., the periodic table of metalloporphyrins was nearly completed [8, 29]. A review article devoted to the porphyrin complexes of individual metals gave examples for metal insertions according to following reactions (1) to (8) [8]; for abbreviations see Table 1. Recent applications and improvements are noted below the respective equations.

2.1.1 Ruthenium

The first ruthenium porphyrins were made by Fleischer et al. [30] however the work required some amendment [31]. A well-defined procedure (Eq. 3) was recommended in a previous review [8]:

$$Ru_2(CO)_9 \xrightarrow[\text{(refluxing benzene)}]{\text{1) } H_2(OEP) \quad \text{2) Py [32]}} RuCO(OEP)Py \tag{3}$$

$Ru_3(CO)_{12}$ in benzene [33], in toluene [34], in acetic or propionic acid [31], or in decalin [35], or $RuCl_3$ in DEGE [36–38] or glycol and formaldehyde [39] have also been used. Tsutsui's protocol using $Ru_3(CO)_{12}$ [33] was frequently mentioned [40–43]. Boiling for 22 to 60 h was necessary in most cases. Only the use of 2-methoxyethanol [44] or decalin [35] as a solvent appears to have been a notable improvement because a reduction of the reaction time to 4 to 5 h in some cases was achieved.

Ruthenium has been fixed to porphyrins also in an "exocyclic", "peripheral" manner. 5-(4-Pyridyl)-10,15,20-tri(*p*-tolyl)porhyrin was treated with $[Ru(NH_3)_5OH_2][PF_6]_2$ at room temperature in acetone for 4 h under conditions which are far too mild for ruthenium insertion to the porphyrin hole. A pentammineruthenium (III) fragment was thus attached to the pyridyl ring yielding a trication which seems to have been isolated as the hexafluorophosphate;

the porphyrin hole then was metallated with a Zn(II) salt which was not specified [45].

2.1.2 *Osmium*

The first osmium porphyrins were obtained by Rohbock [46] (Eq. 4) in diethyleneglycol monomethylether (DEGE).

$$OsO_4 \xrightarrow[\text{(refluxing DEGE)}]{\text{1) } H_2(OEP) \text{ 2) Py [46]}} OsCO(OEP)Py \tag{4}$$

Application of this synthesis to tetraarylporphyrins was less favorable due to the inferior stability of, e.g., OsCO(TTP)L, at high temperatures. Thermal decarbonylation and autoxidation always lead to contamination of the product with OsO_2(TTP) (see Sect. 3.2.2) [47].

Instead of the poisonous OsO_4, $K_2[OsCl_6]$ [48], $[OsCl_2(CO)_3]_2$ [49] and $Os_3(CO)_{12}$ [50] have been used as metal sources. The frequently reported use of $Os_3(CO)_{12}$ is discouraged because this unreactive carbonyl must be added in threefold excess over the porphyrin. Yields of 70% for OEP complexes are reported in an often cited paper [49]. The less soluble the porphyrin, the worse the yield. Starting from H_2(TPP), yields of ca. 40–50% are normal. Both $[OsCl_2(CO)_3]_2$ and K_2 $[OsCl_6]$ are commercially available. The less expensive K_2 $[OsCl_6]$ may be activated prior to osmium insertion with formic acid which is known to produce osmium(II) complexes [51]. The most rapid osmium incorporation (1 h) was achieved with Li_2(TTP) and $[OsCl_2(CO)_3]_2$ in 1,2,4-trichlorobenzene [52]. OsCO(Meso-DME)THF [48] or OsCO(Meso-DME)-EtOH [50] are now available for reconstitution studies with apoglobin.

2.1.3 *Rhodium*

Rhodium was inserted for the first time by Fleischer et al. [53] and its porphyrin chemistry thoroughly investigated especially by Ogoshi et al. (Eq. 5).

$$[RhCl(CO)_2]_2 \xrightarrow[\text{(refluxing benzene)}]{\text{1) } H_2(OEP) \text{ [54–57]}} RhCl(OEP)(H_2O) \tag{5}$$

The unpolar solvent required rather long reaction times, but allowed one to identify a Rh(I) intermediate, $[Rh(CO)_2]_2$(OEP), which carries two $Rh(CO)_2$ groups on both sides on the porphyrin ring and is a typical example of a "bimetallic porphyrin" [8]. A photochemical variant of this insertion in benzene/tetrachloromethane [58] accelerates the reaction and gives a 50% yield of RhCl(OEP)H_2O. Aoyama, Ogoshi et al. [59] used $[RhCl(CO)_2]_2$ and carefully purified benzene. The solution of the porphyrin with the metal carrier was slowly concentrated at room temperature, then heated 2 h to boiling and RhCl(OEP)(H_2O) was obtained in yields of 75 to 80%. Dimeric bis(olefin) rhodium(I) chlorides can replace the Rh(I) carbonyl, e.g. [Rh(COE)$_2$Cl]$_2$ [54, 60]. Wayland [61] et al. recently used a nice variant developed by Grigg et al. [62]. After the treatment of a porphyrin with [RhCl(CO)$_2$]$_2$ in $CHCl_3$ or 1,2-dichloroethane and suspended sodium acetate, addition of iodine to the reaction mixture yielded a iodide RhI(P). For P = TMP or TXP [61], the products were formulated as RhI(P) although they had been purified by chromatography at alumina. However, it is more likely that water is bound as in RhI(Etio-I)(H_2O)$_2$; this sample was characterized by elemental analysis [62].

Dimethylformamide (DMF) has been suggested as a general solvent for metal insertions by Adler et al. [63]. It was used by Gouterman et al. [64] using $RhCl_3(H_2O)_3$ as metal carrier and produced a bisdimethylamine complex, [Rh(Etio-I) (NHMe$_2$)$_2$]$^+$ by decomposition of DMF. Boschi et al. [65–67] performed the reaction of $RhCl_3(H_2O)_3$ with H_2(TPP) or H_2(TTP) in various acetamide and formamide derivatives. Refluxing the reactants for 48 h in pure DMF yielded a salt of the cationic bis(ammine) complex, [Rh(OEP) (NHMe$_2$)$_2$]-[Rh(CO)$_2$Cl$_2$] [65]. From the solvents *N*-methylformamide and *N*-ethylformamide the corresponding cationic bis(ammine) complexes were obtained, while *N*,*N*-dimethylacetamide yielded RhCl(OEP)(NHMe$_2$) or – surprisingly, RhMe(TPP); likewise, ethylrhodium porphyrins RhEt(OEP) or RhEt(TPP) were formed from *N*,*N*-diethylformamide [66]. Diethylacetamide or propionamide produced the acyl complexes RhAc(TPP) or RhCOEt(TPP), respectively [67]. In view of these results, it is doubtful which ligands were obtained in previous experiments using $RhCl_3(H_2O)_3$ in MeOH/DMF (1:4) after refluxing for 20 h; the authors just state that they had obtained RhCl(OEP)L (L being a solvent molecule which appears to be unidentified) [68]. Kadish et al. [69] have also complained about mixtures being obtained in DMF from $RhCl_3(H_2O)_3$ and H_2(TPP).

Because of the lability of the carbonyl group of the acylamides and the stability of the rhodium–ammine bond, the use of $RhCl_3$ with benzonitrile as solvent might be suggested. The use of $RhCl_3$ in glycole in the presence of formaldehyde also appears to be quite practicable [70].

2.1.4 Iridium

The first iridium porphyrins were also made by Fleischer et al. (Eq. 6).

$$\mathrm{IrCl(CO)_3} \xrightarrow[\text{(hot MeOCH}_2\text{CH}_2\text{OH/HOAc)}]{\mathrm{H_2(Hemato\text{-}DEE)}\ [53]} \mathrm{IrCO(Hemato\text{-}DEE)Cl} \tag{6}$$

[Ir(COD)Cl]$_2$ has also been used as a metal carrier. In Kyoto, this insertion was done in refluxing xylene (15 h) and as with Rh, the Ir(I) intermediate [Ir(CO)$_3$]$_2$(OEP) and the desired carbonyliridium(III) complexes IrCl(OEP)-CO were isolated [71]. Sugimoto et al. have extended this work [72].

Representatives of the following Pd-, Pt-, and Ag porphyrins were already known at the culmination of Hans Fischer's work in 1940 [7, 73].

2.1.5 *Palladium*

Palladium is best inserted starting from $PdCl_2$. The solvent benzonitrile was introduced by Eisner [74] who used the complex $PdCl_2(PhCN)_2$ made from $PdCl_2$ and excess PhCN. The complex can be formed in situ and the porphyrin added after dissolution of the $PdCl_2$ (Eq. 7).

$$PdCl_2 \xrightarrow[\text{(refluxing PhCN)}]{H_2(OEP)\ [75]} Pd(OEP) \tag{7}$$

The procedure gives excellent results with both $PdCl_2$ and $PtCl_2$ (see below) [76, 77]. Unsubstituted porphinatopalladium(II) was commercially available for some time [78]. A reaction of 5,10,15,20-tetrakis(4-acetyloxyphenyl)-porphyrin with Pd(II) was noted to occur in aqueous sodium dodecyl sulphate at pH 3.4 and 100 °C. The adduct was stable for 20 h [79].

Collapse of *N*-alkylated porphyrins [80] is an alternative to metal insertion into normal porphyrins. Thus, *N*-benzylprotoporphyrin IX-dimethylester reacts in refluxing methanol with $PdCl_2$ to yield 92% Pd(Proto-DME) within 10 min.

In the course of a study of chlorin and isobacteriochlorin complexes, $PdCl_2$ and DMF were used to prepare Pd(OEP) but partial dehydrogenation occurred in this solvent with the corresponding hydroporphyrins, $H_2(H_2OEP)$ or $H_2(a\text{-}H_4OEP)$ [81]. Therefore, first palladium acetate in $CHCl_3/MeOH$ [82] and then its acetylacetonate in toluene [83] have been applied. In concentrated solutions of the hydroporphyrins, rapid palladium insertion is observed; in dilute solutions, a byproduct is formed which probably contains just partially inserted metal with acetylacetonate or other ligands besides the hydroporphyrin. Earlier reports [84] indeed suggested that metal acetylacetonates would be very suitable for the insertion of divalent metals into sensitive porphinoids, as found [83] with these clean syntheses of $Pd(H_2OEP)$ and $Pd(a\text{-}H_4OEP)$.

2.1.6 *Platinum*

As already stated, benzonitrile is a very useful solvent for the insertion of platinum with $PtCl_2$ (Eq. 8). Since the Pt atom in platinum acetylacetonate is bonded to carbon, it was not used for the insertion of this metal.

$$PtCl_2 \xrightarrow[\text{(refluxing PhCN)}]{H_2(OEP)\ [75]} Pt(OEP) \tag{8}$$

This method was later applied to tetraarylporphyrins as well [76, 85]. $PtCl_2$ in glacial acetic acid seems to be useful to insert Pt(II) into hematoporphyrin-IX

[86]. An intermediate of the metallation procedure, the adduct *cis*-$PtCl_2H_2$-(Hemato-XI), has been characterized [86c]. Tetrabenzporphyrin was transformed into its Pd and Pt complexes, Pd(TBP) or Pt(TBP), with $PdCl_2$ in DMF [87].

2.1.7 Silver

Silver porphyrins are normally prepared from silver(I) acetate. Without precautions, Ag(I) is transformed to Ag(II) in the course of the metallation. Dorough et al. [88] observed two separate steps, Eq. (9a) and (9b) leading first to a disilver(I) porphyrin which collapsed to a silver(II) porphyrin and silver(0) on heating. In an excellent, early study on metalloporphyrins, Haurowitz et al. [89] quantitated the metallic silver that had been formed in the disproportionation reaction (9a, b) applied to H_2(Meso-DME).

$$\text{2AgOAc} \xrightarrow[\text{(warm Py)}]{\text{(a) } H_2\text{(TPP)}} \underset{\text{green}}{Ag_2\text{(TPP)}} \xrightarrow[\text{(boiling)}]{\text{(b) Py}} \underset{\text{reddish-orange}}{\text{Ag(TPP)} + \text{Ag}} \quad (9)$$

If AgOAc is reacted with porphyrins in glacial acetic acid, Ag(II) porphyrins are formed directly [88, 90]. Whether this is due to oxidation by protons [88], or more likely, by aerial dioxygen, is an open question. Anyway, the formation of the Ag(II) porphyrin even occurs when AgOAc is used in just 1.2 molar excess to the porphyrin in acetonitrile [91]. Instead of pyridine, dimethyl formamide [63] is frequently used as a solvent for the preparation of Ag(II) porphyrins [92–95]. With certain water-soluble porphyrin ligands, the disproportionation reaction (9a) is retarded, and the corresponding disilver(I) porphyrins can be studied in solution [96].

2.1.8 Gold

Gold porphyrins were first obtained by Rothemund et al. [97] in glacial acetic acid (Eq. 10).

$$\text{K[AuCl}_4\text{]} \xrightarrow[\text{(refluxing HOAc/NaOAc)}]{H_2\text{(TPP) [98]}} \text{[Au(TPP)][AuCl}_4\text{]} \quad (10)$$

The method was checked for the synthesis of [Au(Meso-DME)]OAc and [Au(Etio)][$AuCl_4$] [99]. The $[AuCl_4]^-$ ion was destroyed by cathodic electrolysis and thus exchanged for $[ClO_4]^-$. AuCl(TPP) can also be made (see Sect. 2.5).

2.2 *Metal Complexes of Carbon-Based Tetrapyrroles Other Than Metal Porphyrins*

Not only porphyrins, but tetrapyrrole ligands in general are interesting partners for the coordination of noble metals. This section will refer to the synthesis of some tetrapyrrole complexes other than porphyrins; from time to time, derivatives of these ligands will be mentioned in the later sections on the reactions of noble metal porphyrins because most of these tetrapyrrole complexes behave similarly as regards their axial coordination chemistry.

Hydroporphyrins – Palladium complexes of hydroporphyrins, i.e. chlorins and isobacteriochlorins, have already been mentioned in the previous section. In principle, the insertion of noble metals into hydroporphyrins is difficult, because at the high temperatures required, the metal carriers tend to decompose in some way to produce noble metals themselves or derivatives which could act as heterogeneous or homogeneous dehydrogenation catalysts, thus a dehydrogenation of the hydroporphyrin to the corresponding porphyrin may occur along or dominate over metal insertion. This problem may be circumvented by using hydroporphyrins in which alkyl groups replace the hydrogen atoms that could be removed in dehydrogenation reactions, or by generating the hydroporphyrin chromophore by reductive alkylation. Thus, an osmium(II) complex $OsCO(OEPMe_2)Py$ (**4**) containing a porphodimethene ligand, 5,15-dimethyl-5,15-*H*-octaethylporphyrinate, was made by reductive methylation of OsCO-(OEP)Py with sodium anthracenide/methyl iodide [100].

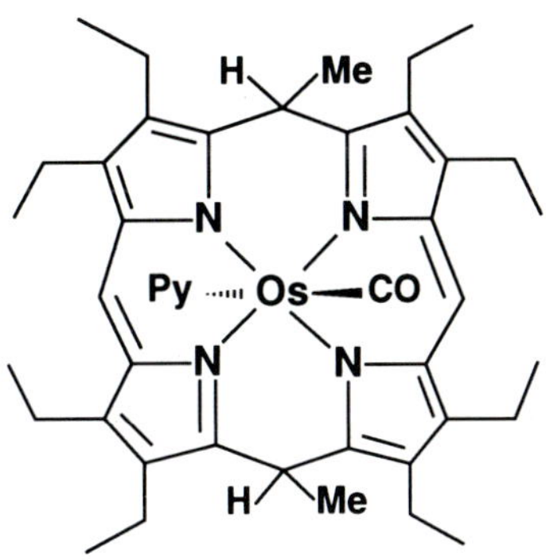

4: $OsCO(OEPMe_2)Py$

Porphyrin analogs, homologs, and porphyrinoids – The class of tetrapyrrole chromophores can be enlargened by alteration of the numbers of carbon atoms linking the pyrrole rings [101]. A beautiful porphyrin isomer, porphycene, has been prepared by Vogel and coworkers [102]. This porphyrin isomer, $H_2(Pyc)$, and its tetrapropyl derivative, H_2(TPrPyc), have been transformed into their palladium and platinum complexes [103], **5a** and **5b**, respectively; the corresponding ruthenium and osmium complexes, **5c** to **5e**, have also been obtained from the respective metal carbonyls $M_3(CO)_{12}$ in decalin or DEGE [104]. It is amazing how rapidly these metallations occurred. Possibly at higher temperatures, the porphycene chromophore opens up more easily to metal insertion. The smaller hole does not cause any problem to the accomodation of the "soft acids" which are the Ru(CO) and Os(CO) fragments. OsCO(TPrPyc)Py (**5d**) was transformed in the dioxoosmium(VI) complex, OsO_2(TPrPyc) (**5e**) with *m*-chloroperbenzoic acid.

5: MLL′ (TPrPyc)

No	M	L	L′
5a	Pd	–	–
5b	Pt	–	–
5c	Ru	CO	Py
5d	Os	CO	Py
5e	Os	O	O

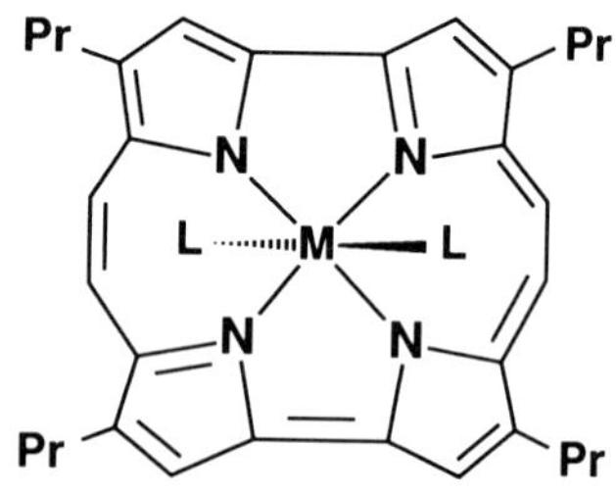

5: MLL′(TPrPyc)

Corroles and corrinoids – For results on the noble metal complexes of this interesting class of tetrapyrroles, see the article by S. Licoccia in this volume [105].

2.3 Noble Metal Azaporphyrinates

Phthalocyanines – Pthalocyanine complexes form a very large class among tetrapyrrole complexes. Their chemistry, physics, and applications for novel materials are being reviewed in a multivolume series [106]. Usually they are made by reductive tetramerization of phthalodinitrile with a metal salt (see below), but metallations of free phthalocyanine, H_2(Pc), are also documented. Work in the last two decades has been concentrated on phthalocyanine complexes of Ru, Os, Rh, Pd, Pt, and Ag; Berezin gives references on IrCl(Pc) and AuCl(Pc) [107].

Ruthenium and Osmium Phthalocyaninates – By cyclization of phthalodinitrile with $RuCl_3(H_2O)_3$ or OsO_4 in the presence of carbon monoxide or with the corresponding trimetal dodecacarbonyls, the pure complexes RuCO(Pc) Py and OsCO(Pc) Py have been obtained in good yields [108]. *o*-Cyanobenzamide and $RuCl_3(H_2O)_3$ [109] produced so-called "crude Ru(Pc)" and, after extraction with pyridines, complexes like Ru(Pc)Py_2.

A series of publications by Hanack et al. [110] is devoted to the synthesis of metal phthalocyanine systems in which bridging axial ligands serve to construct metallophthalocyanine stacks. These products are studied as novel electrical materials. Therefore, the synthesis of Ru(II) and Os(II) phthalocyanines, their axial ligation with monodentate ligands, and interaction with bidentate ligands have been investigated in depth. Syntheses of these compounds started from the "crude Ru(Pc)" mentioned above which was transformed into the bis(dimethylsulfoxide) complex, $Ru(Pc)(DMSO)_2 * 2DMSO$, on heating in DMSO, and from Ru(Pc) which was obtained in a pure form after slowly heating Ru(Pc) $(DMSO)_2 * 2DMSO$ to 330 °C in vacuo [111]. Easy procedures for the preparation of soluble octakis(pentyloxy)- or tetrakis(*tert*-butyl) phthalocyaninatoruthenium(II), and Ru(Pc) were reported recently [112]. $RuCl_3$ in ethoxyethanol served to metallate a preformed phthalocyanine.

Heating $OsCl_3 * H_2O$ with *o*-cyanobenzamide in a naphthalene melt gave a product, $OsL_2(Pc)$ with unidentified axial ligands L, which was slowly heated to 400 °C in vacuo whereupon analytically pure Os(Pc) remained [113]. Extraction of the product with pyridine before sublimation yielded $OsPy_2(Pc)$ which loses two moles of pyridine on heating in vacuo to 400 °C as above [114].

By analogy to the corresponding porphyrins (see Sect. 3.2), the species M(Pc) (M = Ru, Os) whenever formed in the absence of donor ligands ought to form dimers $[M(Pc)]_2$ with metal–metal double bonds. Indeed, this appears to be the case [115]. Amorphous paramagnetic $[Ru(Pc)]_2$ was identified by large-angle X-ray scattering and magnetic measurements [116]. On average, six dimeric units closely approach one another and are stacked in a monodimensional array. $[Ru(P)]_2$ is stable in the solid state, but is autoxidized in THF.

Homborg et al. used K_2OsCl_6 [117] for the formation of an Os(III)(Pc) system from phthalodinitrile and iodine which they formulate as an acid, $H[OsCl_2(Pc)]$ probably containing some water; the proton may be located at one of the *meso* nitrogen atoms [117b]. The anion of this acid can be obtained in a tetrabutylammonium salt, $[NBu_4][OsCl_2(Pc)]$, which is well-characterized by resonance Raman spectra [117c]. The corresponding Ru compounds have also been made.

Rhodium phthalocyaninates – Well-defined rhodium(III) phthalocyanines RhX(Pc)(MeOH) (X = Cl, Br, I) with halide and methanol as axial ligands were prepared for photochemical investigations [118].

For the insertion of noble metal ions into phthalocyanine $H_2(Pc)$, the latter is favorably activated by transformation into its lithium derivative. Thus, reaction of $Li_2(Pc)$ with $[Rh(COD)Cl]_2$ in a large excess of a donor solvent L yielded the Rh(II) complexes, Rh(Pc) L_2 (L = Py, $BuNH_2$, Bipy, Pyz) [119]. The formation of these Rh(II) species is very remarkable because Rh insertions normally yield either Rh(I) or Rh(III) derivatives. Recently, tris(ethylenediamine)-rhodium(III) iodide $[Rh(en)_3]I_3$ and phthalodinitrile have been used to prepare $H[RhI_2(Pc)] * 2$phthalodinitrile. This material was then transformed into $[Rh(OH)_2(Pc)]^-$ by treatment with first H_2SO_4 and then KOH. Treatment of the dihydroxoanion with a variety of acids HX or anions X^- produced the corresponding rhodium(III) porphyrins $[RhX_2(Pc)]^-$ (X = Cl, Br, I, N_3, CN, NCO, NCS, SeCN, PhO, PhS, PhSe) which were isolated as tetrabutylammonium salts and characterized by resonance Raman spectra. Diacidorhodates(III) and species with O, S, and Se donor ligands have continued these series [120].

Palladium, platinum, and silver pthalocyaninates – 5,6-Di-substituted isoindoles were condensed in the presence of $Pd(acac)_2$ or $PtCl_2$ to obtain soluble octa-alkoxy-substituted phthalocyaninato palladium(II) and platinum(II) complexes [121]. Tetrakis-(neopentyloxy) phthalocyaninatosilver(II) was obtained in high yield from $AgNO_3$ and the respective metal free phthalocyanine in DMF at 75 °C [122].

5,10,15,20-Tetraaza-2,3,7,8,12,13,17,18-octaethylporphyrin – This tetraazaporphyrin, H_2(OETAP), was metallated with $[Rh(CO)_2Cl]_2$ as usual and isolated as RhI(OETAP) [123].

Diazaporphyrins – Like the phthalocyanines, a 10,20-diaza-porphinoid palladium complex was composed by cyclooligomerization. The condensation of two suitable bispyrrole halves yielded the tetrapyrrole complex, *cis* [1,11-dimethoxy-2,2,3,3,7,7,8,8,12,12,13,13,17,17,18,18-hexadecamethyl-10,20-diaza-decahydroporphyrinato]palladium(II) [124].

2.4 Water-Soluble Porphyrin Complexes

The syntheses, and especially the purification, of water-soluble porphyrin complexes are often different from the above-mentioned methods, and experimentally more difficult, because of the salt-like character of the compounds. Thus, besides the reaction conditions, some notes on the workup are given. Some papers do not specify the counter ions neutralizing the charge of the respective metalloporphyrin. In such cases, just the respective ion of the porphyrin complex is noted. The porphyrin systems involved are specified in Table 1.

Ruthenium – Using $Ru_3(CO)_{12}$ as metal carrier, compounds designated as $[RuCO(TPPS_4)]^{4-}$ or $RuCO(TH_4CPP)$ have been obtained in DMF [125] or DEGE [126]. "$RuCO(TH_4CPP)$" was isolated from pyridine, thus it might have had Py as another axial ligand. The salt $[Ru(TEAP)Py_2]$ $[SO_4]_2$ was prepared from $RuCl_3$ in water and isolated after addition of pyridine [127].

For the preparation of an uroporphyrin-I derivative of the composition RuCO(Uro-I)H_2O, Uro-I octamethylester was metallated with $Ru_3(CO)_{12}$ in xylene and saponified with NaOH [128].

Osmium – In contrast to ruthenium it was not possible to insert osmium into a water-soluble porphyrin. The metal carrier (K_2OsCl_6 or Os_3CO_{12}) seemed to be deactivated by the polar groups of the porphyrin. Therefore OsCO(TPP)THF, OsO_2(TPP), and OsCO(TMeCPP)THF were prepared and transformed into Na_4[OsCO($TPPS_4$)H_2O], Na_4[OsO_2($TPPS_4$)], and OsCO(TH_4CPP)THF, respectively, by sulfonation or saponification [51]. The salts have been purified by ultrafiltration and their anions identified by electrophoresis [129].

Rhodium – The reaction of $[Rh(CO)_2Cl]_2$ with $Na_4[H_2(TPPS_4)]$ in methanol, subsequent oxidation of the resulting Rh(I) complex with H_2O_2, extraction with methanol, and precipitation with acetone (or taking an aqueous solution to dryness) yielded a hydrated rhodium(III) porphyrin salt, $Na_3[Rh(TPPS_4)(H_2O)_2] * xH_2O$. with x = 14 [130], 21 [131] or 22 [132]. $RhCl_3 * (H_2O)_3$ dissolved in DMF may also be used as metal carrier [68, 132]. The product was best purified by chromatography on a water-washed alumina column using 0.02 n NaOH as eluent and further on a Sephadex G-10 column to remove extraneous ions [132]. These rhodium porphyrin ions have been used for the study of metalloporphyrin-ligand equilibria and photochemistry [68, 129, 133].

Iridium – The authors are not aware of any water-soluble iridium porphyrin.

Palladium – Palladium was inserted starting from a solution of $[NEt_4]_4$-$[H_2(TPPS_4)]$ in methanol/benzonitrile (2:1) adding $PdCl_2$ dissolved in benzonitrile. The product $[NEt_4]_4$ $[Pd(TPPS_4)]$ was purified by HPLC using a reversed phase column [134]. For photophysical studies, $[Pd(TPPS_4)]^{4-}$, Pd(TH_4CPP) and $[Pd(TMPyP)]^{4+}$ were made from $PdCl_2$ in DMF [135]. The products were purified on Dowex or Sephadex columns but no information was given on analytical data.

Platinum – Pasternack reported the synthesis of a water-soluble platinum porphyrin using $[H_2(TMPyP)]$ $[tosylate]_4$ and $[cis\text{-}Pt(H_2O)_2(DMSO)_2]^{2+}$ as metal carrier [136]. The product, $[Pt(TMPyP)][PF_6]_4$, was precipitated by addition of NH_4PF_6 to a solution of the crude product in water. It was used for intercalation studies with single-stranded poly(deoxyadenylic acid). Na_4-$[Pt(TPPS_4)]$ was prepared by prolonged heating of $Na_4[H_2(TPPS_4)]$ with $Pt(acac)_2$ and sodium acetate in glacial acetic acid [137]. While a reproduction of this latter method required sodium carbonate instead of sodium acetate [138], the reaction of $[cis\text{-}Pt(H_2O)_2(DMSO)_2]^{2+}$ with $Na_4[H_2(TPPS_4)]$ in chloride-free water at pH = 7 (100°C, 18 h) as described by Pasternack [136] is more elegant and gives a quantitative yield of $Na_4[Pt(TPPS_4)]$ [138] after ultrafiltration [51].

Silver – $Na_4[Ag(TPPS_4)]$ was obtained by stirring a suspension of Ag_2O in a solution of $Na_4[H_2(TPPS_4)]$ in water at room temperature. The excess metal carrier was separated from the product by filtration [139]. Using AgOAc for the incorporation, the purification of the paramagnetic $Na_4[Ag(TPPS_4)]$ was more difficult. It was isolated as a decahydrate and was oxidized to a diamagnetic Ag(III) compound, Na_3 $[Ag(TPPS_4] * 6H_2O$. These compounds were used for demetalation studies [140]. The cation $[Ag(TTMAP)]^{4+}$ was prepared from $AgNO_3$ in water and precipitated as a perchlorate salt [141].

Gold – Water-soluble gold porphyrin ions, $[Au(TPPS_4)]^{3-}$ and $[Au(TMPyP)]^{5+}$ (as the 3- and 4-pyridyl isomers), were obtained from the respective free-base porphyrins in a mixture of water/pyridine/lithium chloride containing $KAuCl_4$ [142]. The products were purified by extraction with acetone and methanol and subsequent HPLC, however, the counterions were not determined.

2.5 Structure of Noble Metal Porphyrins

As already mentioned, the coordination type of the Ru, Os, Rh and Ir porphyrins is **C** (Fig. 2), i.e. distorted octahedral about the metal which is in most cases a d^6 system; occasionally, d^5, d^4, or d^2 systems are encountered. Table 2 gives a compilation of compounds the molecular structures of which have been determined by X-ray crystallography. The numbers of entries given for the individual noble metals approximately reflects the intensity of research done for the respective metal. For details and special structural aspects, the reader is referred to the original literature and a previous review by Scheidt and Lee [143]. Here, just a few general notes will be made. Phthalocyanine systems are not incorporated.

Only a few compounds containing Ru, Os, Rh or Ir do not possess type **C**. Notable are the metalloporphyrin dimers with Ru = Ru (entry 3) or Rh–In bonds (entry 34), the alkyl or acyl rhodium(III) (entries 31, 36, 38), alkyliridium(III) (entry 39) or arylruthenium(III) compounds (entry 12). All these species contain pentacoordinate noble metal ions, Type **B**. A reason for this unusual behavior of the metal ions is seen in the strong *trans* effect of the axial ligand L in these systems which is a strong σ-donor (metalloporphyrinyl, alkyl, aryl, acyl) which precludes further coordination in the *trans* position of L. The reluctance of RhMe(OEP) to accept an axial ligand is shown by its crystallization as a π-dimer from *n*-hexane in presence of the base 1-Meim.

An exception is found with alkyl iridium porphyrins and π-acceptor ligands. In the *n*-propyl iridium complex, $Ir(C_3H_7)(OEP)L$ with $L = PPh_3$ (entry 41) and DMSO, type **C** is found. The sulfoxide is S-bound in the DMSO complex.

AuCl(TPP) (entry 50) also belongs to coordination type **B**, with a rather long Au–Cl bond and a hydrogen bridge to the $CHCl_3$ molecule; the gold(III) ion is practically in the porphyrin plane to maintain good π-bonding to the porphyrin ligand.

The bidentate bis(diphenylphosphino) alkane ligands (entries 8 and 42) are not capable of forcing the porphyrin ligand to leave the equatorial position of noble metal porphyrins. Either two bidentates are bound in a monodentate fashion, or a bidentate bridges two metalloporphyrin entities.

Table 2. Representative precious metal porphyrins as studied by X-ray crystallography

Entry	Complex	Ref.
a) Ruthenium porphyrins		
1	RuCO(TPP)EtOH	[144]
2	RuCO(TPP)Py	[145]
3	$[Ru(P)]_2$	[146]
4	$[RuOH(P)]_2O$	[147]
5	$[RuCl(P)]_2O$	[148]
6	$RuBr(OEP)PPh_3$	[149]
7	RuOEt(TPP)EtOH	[150]
8	$Ru(OEP)\ (dpm)_2$ [a]	[151]
9	RuCO(PBP)Py [b]	[152]
10	$[RuPh(NPhOEP)]BF_4$ [c]	[153]
11	$Ru(TMP)N_2(THF)$	[154]
12	RuPh(OEP)	[155]
13	$RuCO(TCl_8PP)$ (StOx) [d]	[156]
14	$Ru(TPP)Br_2$	[157]
15	RuCO(OEP)TEMPO [e]	[158]
16	$Ru(OEP)(SPh_2)_2$	[159]
17	$Ru(TPP)(CO)_2$	[160]
b) Osmium porphyrins		
18	$OsCO(OEPMe_2)Py$	[100]
19	$[OsOMe(OEP)]_2O$	[161]
20	$Os(OR)_2(TPP)$ [f]	[162]
21	$OsO_2(OEP)$	[163]
22	$OsO_2(TPP)$	[164]
23	$Os(NC_6H_4NO_2)_2(TTP)$ [g]	[165]
24	$Os(OEP)\ (OPPh_3)_2$	[166]
25	$Os(TPP)\ (OPPh_3)_2$	[166]
26	$Os(SiEt_2THF)$ (TPP) THF	[167]
27	$Os(SC_6HF_4H)_2(TTP)$	[168]
c) Rhodium porphyrins		
28	$[Rh(NHMe)_2(Etio)]Cl$	[64]
29	$[Rh(NHMe)_2(TPP)]Cl$	[169]
30	RhCOPh(Etio)	[170]
31	RhMe(OEP)	[171, 172]
32	$[Rh(CO)_2](OEP)$	[173]
33	$RhCl(OEP)PPh_3$	[60]
34	(OEP)RhIn(OEP)	[174]
35	$[RhCR_2(TPP)(CNCH_2Ph)]PF_6$ [h]	[175]
36	Rh(COOEt)(OEP)	[176]
37	$RhCl_2(STPP)$ [i]	[177]
38	RhMe(OETAP)	[123]
d) Iridium porphyrins		
39	$Ir(OEP)\ (C_8H_{13})$ [j]	[178]
40	IrCl(OEP)CO	[179]
41	$Ir(OEP)(C_3H_7)(PPh_3)$	[180]
42	$[Ir(OEP)Cl]_2(dppe)$ [k]	[181]
43	IrCO(TDBPP)Cl [l]	[182]

Table 2. Continued

e) Palladium and platinum porphyrins		
44	Pd(TPP)	[183]
45	Pd(OEP), Pd(H_2OEP)	[183]
46	Pd(TDBOHPP)[m]	[185]
47	Pd(SDPDTP)[n]	[186]
48	Pt(TPP)	[187]
f) Silver and gold porphyrins		
49	Ag(TPP)	[188]
50	AuCl(TPP) * $CHCl_3$	[189]

[a] dpm: Bis(diphenylphosphino)methane
[b] H_2 (PBP): "picnic-basket" porphyrin
[c] H(NPhOEP): *N*-Phenyloctaethylporphyrin
[d] StOx: styrene oxide.
[e] TEMPO: 2,2,6,6-Tetramethylpiperidin-1-oxyl
[f] R = OEt, *i*Pr, OPh
[g] Bis(arylimido)osmium(VI)porphyrin
[h] R: -NHCH$_2$Ph
[i] H(STPP): Tetraphenyl-21-thiaporphyrin
[j] C_8H_{13}: *cis*-bicyclo(3.3.0)oct-1-yl
[k] (dppe): 1,2-Bis(diphenylphosphino)ethane
[l] H_2(TDBPP): 5,10,15,20-Tetrakis(3,5-*tert*-butylphenyl)porphyrin
[m] H_2(TDBOHPP): 5,10,15,20-Tetrakis(3,5-*tert*-butyl-4-hydroxyphenyl) porphyrin
[n] H(SDPDTP): 5,20-Diphenyl-10,15-ditolyl-21-thiaporphyrin

The d^8 and d^9 metal ions, Pd(II), Pt(II), and Ag(II), as in other complexes, prefer the square-planar coordination type given as **A** in Fig. 2 (entries 44–50).

Since all the noble metal ions are "soft" Lewis acids, their size is adaptable according their π-donor or π-acceptor capacity. Thus, irrespective of their oxidation state, all the noble metal ions fit well into the central hole of the porphyrinato ligand, and, generally, the $C_{20}N_4$ core of the porphyrin ring is nicely planar.

A notable example is the comparison of the osmium derivatives (entries 18 to 27) in which the porphyrin rings all seem to be planar, and the osmium always sits in the N_4 plane of the porphyrin ring, irrespective of the oxidation states between 2 and 6 which are encountered in the Os ions involved.

An interesting compound is (*N*-phenyloctaethylporphyrinato) phenylruthenium(II) (entry 10) in which an agostic hydrogen of the *N*-phenyl group completes the heavily distored octahedral coordination sphere of the Ru(II) ion. The Ru ion protrudes by 14 pm towards the axial phenyl anion.

In the Rh (entry 37) or Pd thiaporphyrin derivatives (entry 47), the thiapyrrole-tripyrrole macrocycle is not planar. Because the thiaporphyrin only carries one central proton, anions are required as axial ligands in its chelates to neutralize the latter. Pd(SDPDTP) (entry 47) does not have an anion. It has been formed by one-electron reduction from Pd(SDPDTP)Cl. An extra electron resides in the thiaporphyrin π-system, the Pd ion stays divalent and square planar.

3 Inorganic Reactions

3.1 General Remarks

The usual products obtained from metal insertion reactions are shown in Table 3. They are used as starting materials for any axial ligand substitution processes.

The iron and cobalt homologs produce the rather stable d^6 complexes with the distorted octahedral coordination type **C**, the nickel and copper homologs are square planar d^8 or, for Ag, d^9 systems of type **A**. The complexes of type **C** have a rich substitution chemistry, very frequently accompanied with redox reactions, while the complexes of type **A**, due to the reluctance of the d^8 or d^9 systems to adding axial ligands, are only subject to redox reactions. The inorganic chemistry of the noble metal porphyrins is presented groupwise; because of the richness in compounds for Ru and Os, there are sections "axial ligand substitution" with constant oxidation state at the metal (Sect. 3.2.1), and "oxidation-reduction reactions", frequently accompanied with axial ligand substitution processes (Sect. 3.2.2).

For brevity, the material will be presented in reaction schemes and tables. As far as possible, the dyads Ru/Os, Rh/Ir, Pd/Pt, and Ag/Au will be treated together. The homology of these pairs is expected due to the fact that the size of the two respective ions is very similar (lanthanoid contraction!).

3.2 Ruthenium and Osmium Porphyrins

3.2.1 Axial Ligand Exchange at Ru(II) and Os(II) Species

Ru(II) and Os(II) porphyrins with small molecules as axial ligands – The carbonylmetal(II) species MCO(P)L′ (M = Ru, Os; L′ = H_2O, ROH, THF) emerging from the metal insertions and subsequent chromatographic purifications are

Table 3. Products from metal insertions (for abbreviations, see Table 1)

Noble metal	Compounds 4d-metal	5d-metal	Metal oxidation state	Coordination type
Ru, Os	RuCO(P)THF	OsCO(P)THF	+ 2	**C**
Rh, Ir	$RhCl(P)H_2O$	IrCl(P)CO	+ 3	**C**
Pd, Pt	Pd(P)	Pt(P)	+ 2	**A**
Ag, Au	Ag(P)	$[Au(P)]^+$	+ 2, + 3	**A**

rather stable; however, the neutral donor ligand L′ *trans* to CO is labile. ^{1}H NMR experiments demonstrated, e.g., that in RuCO(*i*PrTPP)(4-*t*Bupy) the axial ligand 4-*t*Bupy at elevated temperatures is exchanged rapidly with excess 4-*t*Bupy in solution [40a].

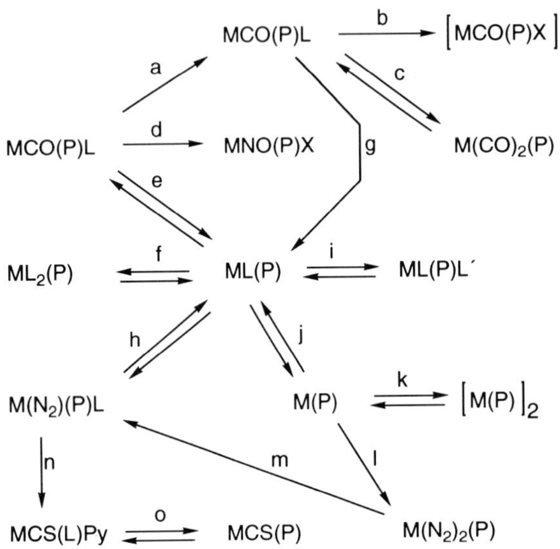

Scheme 1. Reaction paths a–o of ruthenium (II) and osmium (II) prophyrins starting from carbonylmetal (II) complexes MCO (P) L′ (M = Ru, Os). For reaction conditions and references, see Table 4

The substitution chemistry of the Ru(II) and Os(II) porphyrins is synoptically presented in Scheme 1 and Table 4. The lability of the *trans*-L′ in MCO(P)L′ may be used to prepare a large variety of complexes MCO(P) L with other ligands L where the donor atom may be any oxygen-, nitrogen-, or sulfur donor (reaction path a of Scheme 1 and Table 4), cyanide (paths a, b) or carbon monoxide (c) [190]. Coordinated cyanide in $[OsCO(OEP)CN]^-$ can be transformed into isonitrile with methylfluorosulfonate [191].

With donor atoms that belong to strong π-acceptors, e.g. NO, NO^+ or $P(OMe)_3$, a strong *trans*-labilizing effect is exerted upon the original CO ligand which is then easily expelled (path c: (X = NO)) to yield the dinitrosyls $M(NO)_2(P)$ as primary products of the reaction with NO; a *trans*-labilizing effect of the two NO groups is prevented by their disproportionation, one of them being bound as NO^+, the other one as NO^-. The latter is then replaced by alkoxide in the presence of alcohols. The resulting complexes MNO(P)X (X = OR) are then stabilized by a push-pull-effect. The origin of these *trans*-effects has been discussed previously at length [190] and will not be repeated

Table 4. Details for the reaction paths a–o of Ru(II) and Os(II) porphyrins depicted in Scheme 1. Negative sign means reverse reaction

Path	M = Ru reactant	Conditions	Ref.	M = Os ractant	Conditions	Ref.
a	Bupy	L′ = H_2O, EtOH	[40a]	Py = L	P = OEP	[46]
a	HIm		[192]	PPh_3	P = OEP	[46]
b	SR^-	X = SR^-	[193]	CN^- = X^-	X = CN^-	[202]
c	CO	L = CO	[194]	CO	L = CO	[47]
d	NO	MeOH; L′ = Py	[195]	NO	X = OMe, F	[196]
	NOX	X = Cl, Br	[39]	NO	X = NO C_6H_6	[196]
e, f	Py, DMSO	Photolysis	[32]	Py	Photolysis	[43,49,197]
a, g, f	dppe	Room temp	[151]	$P(OMe)_3$	CH_2Cl_2	[46]
− f	PBu_3, Vac	P = TMP, OEP	[198]			
a, g, f	C=N*t*Bu	P = $TpCF_3PP$	[199]			
	Vacuum	L = PBu_3	[198]			
a, g, f	$AsPh_3$, $P(OR)_3$		[39]			
a, g, f	$CNCH_2CF_3$ = L		[200]			
− f, − g	CO, L = DMF, PhCN		[201]	L = $P(OMe)_3$		[202]
− f, i				Py = L′	L = THT	[203]
− f, − j	Vacuum	L = MeCN	[204]			
− f, − j, k	Vacuum	190–210 °C	[146]	Vacuum	190–210 °	[49]
l	2 N_2		[204]			
− h	THF = L	N_2, 20 °C	[204]			
m, h	THF = L′ = L		[154]			
− h, i	L′ = O_2	P = “picnic”	[205]			
h, f				NH_3 = L	THF/N_2H_4	[206]
h, f				SR_2 = L	L = THF	[203]
n				$CSCl_2$; Py	NaHg/THF	[207]
h, f				THF = L	L = THF	[206]
n, o				CS_2, Py	L = THF	[207]
− k, − j, f	SR_2	Room temp.	[159]			
a, g	PF_3 = L	L = CO	[208]			
m, h	L = C_2H_4	P = TMP	[209]			

here. The formation of bis(phosphite) complexes can be explained on similar terms (paths a, g, f; L = $P(OMe)_3$).

PPh_3 behaves differently for M = Ru or Os. In RuCO(P)L′, the CO ligand is less firmly bound as in OsCO(P)L′. Therefore, the *trans*-effect of PPh_3 suffices to expel the CO ligand in RuCO(P)L′ (paths a, g, f), but not in OsCO(P)L′ which can be transformed into OsCO(P) PPh_3 (path a).

The fact that two π-acceptors in *trans*-position labilize each other is seen in the lability of the dicarbonyl and bis(trimethylphosphite) complexes. While OsCO(OEP)$P(OMe)_3$ cannot be formed from OsCO(OEP)L′, it is obtained by bubbling CO through a hot CH_2Cl_2 solution of Os(OEP)$[P(OMe)_3]_2$ for 30 min (− f, − g).

Inertness of hexacoordinate Ru(II) or Os(II) porphyrins – In contrast to the systems in which L is a strong π-acceptor, the strongly basic bis-nitrogen donor systems (L = Py, 1-Meim) are substitutionally inert [210]. From MCO(P)L′, they can be prepared by photolytic ejection of the CO ligand in excess L (paths e, f), but, especially with osmium, photolysis is rather ineffective. Therefore, a deviation was made via the oxidation of OsCO(P)L′ to OsO_2(P) (see Sect. 3.2.2, Scheme 2, path d, Table 5) and reduction of this species with hydrazine hydrate to the dinitrogen complex, OsN_2(P)L′. This very labile dinitrogen complex reacts rapidly with nitrogen or sulfur donors, e.g. NH_3, Py, 1-Meim, or THT (paths h, f).

Because of their structural and spectroscopic analogies with the hemochromes Fe(P)L_2, e.g. the protoporphyrin derivative **3** (L, L′ = Py or 1-Meim), the corresponding 4d and 5d homologs are named "ruthenochromes" or "osmochromes". The hemochromes derive their name from the cytochromes, the widespread electron-carrying heme proteins. Cytochrome b (coordination type **F**, M = Fe) has two imidazole donors from histidine side chains at the central iron, cytochrome c (coordination type **G**, M = Fe) an imidazole and a methylthioether function from a methionine. **F** is an axially symmetrical, **G** an axially unsymmetrical system.

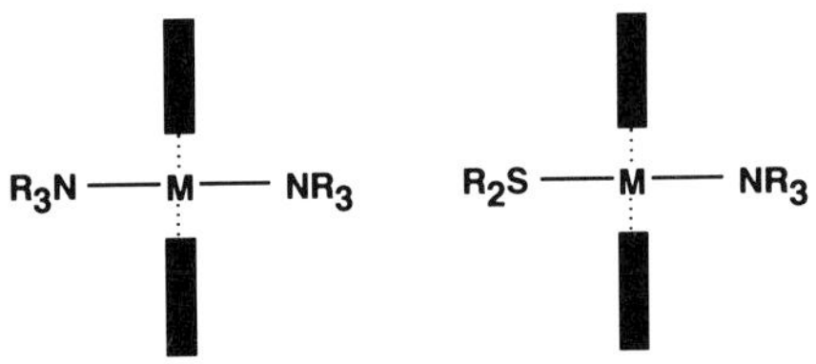

F G

Fig. 3. Axial system of cytochrome b (type **F**) and of cytochrome c (type **G**)

While the hemochromes are substitutionally labile, their autoxidation ultimately producing μ-oxo complexes of the type $[Fe_2(P)_2]O$ accompanied with the loss of the nitrogen donor axial ligands, the ruthenochromes and osmochromes are stable. Autoxidation occurs in an outer-sphere electron transfer, e.g. from Os(II) to a dioxygen molecule, see Sect. 3.22.

The photochemically initiated substitution reactions depicted in Scheme 1 (paths e, f, i) can also be performed with Ru(Pc) systems [211].

Possible analogies to heme – The reconstitution of apomyoglobin with its prosthetic group, the heme [Fe(Proto)L_2], under the exclusion of oxygen, restores deoxymyoglobin [Fe(Proto)G] (G = Globin) and can be viewed as an axial ligand substitution of an iron(II) porphyrin Fe(P)L_2, two labile ligands L being replaced by an imidazole nitrogen of the proximal histidine. In fact, due to the instability of these Fe(II) porphyrins in aerobic aqeuous solutions, in which they are transformed into μ-oxo complexes of the type $[Fe_2(P)_2]O$ ("hematins"), reconstitutions are done at the insensitive Fe(III) stage [212]. On exposure of suitable hemins Fe(P)Cl to apoglobin solutions, the corresponding metmyoglobins [Fe(Proto)G(OH)] are formed which then are reduced to the Fe(II) state, deoxymyo-globin. With the synthesis of Ru(II) and Os(II) porphyrins at hand, the question arose whether these porphyrins could be reconstituted with apoglobin to produce "ruthenoglobin" or "osmoglobin".

The drastic conditions of the metal insertions preclude the preparation of the protoporphyrin derivatives; it is to be expected that the labile vinyl groups of this porphyrin will not survive the metallation procedures. Therefore, the mesoporphyrin derivatives, RuCO(Meso-DtBuE)EtOH [213, 214] and OsCO-(Meso-DME)THF [48, 50] were prepared and their ester groups saponified. Reconstitutions were done with apomyoglobin or apohemoglobin, which produced "ruthenoglobin" [Ru(Meso)G] [213, 214] from the ruthenium(II) complex Ru(Meso) $(DMF)_2$ and "carbonylosmoglobins" [OsCO(Meso)G] [50, 215] from carbonylosmium(II) complexes OsCO(Meso)L, respectively.

[Ru(Meso)G] reacts slowly with CO and O_2 yielding [RuCO(Meso) G] and $[Ru(Meso)G]^+$, respectively; contrary to Ru(Meso) $(DMF)_2$, which is said to bind dioxygen reversibly at 0 °C, ruthenoglobin is simply autoxidized. [RuCO(Meso)G] can be formed both by reconstitution of apomyoglobin with RuCO(Meso) (H_2O) and by carbonylation of [Ru(Meso)G] [213]. While [FeCO(Proto)G] is quantitatively and rapidly photolyzed, photolysis of [RuCO(Meso)G] is only achieved above 40 °C and occurs only in yields below 50%. It seems that, contrary to deoxymyoglobin [Fe(Meso)G], [Ru(Meso)G] is hexacoordinate (in the beginning of the type [RuL(Meso)G] with L = DMF or H_2O), then ages rapidly and finally occurs as a "denatured" protein in which both axial ligand positions at the Ru(II) ion are occupied by donor atoms of the protein side chains (proximal and distal histidine, a "hemochrome"–like situation).

It was not surprising that [OsCO(Meso)G] did not change on illumination at room temperature [215]. Other reports on the photolytic decarbonylation and oxygenation rely only on visible absorption spectra. The spectra of the oxygenated Ru and Os species look like "hemichrome" spectra. Therefore, it seems more likely, that "ruthenichromes" or "osmichromes" are formed on exposure of these Ru(II) and Os(II) porphyrins to dioxygen (see Sect. 3.2).

The comparison of Fe(II) with Ru(II) and Os(II) in myoglobin then discloses the necessity of a labile sixth coordination site for an effective reversible dioxygen binding. As the heavier (4d, 5d) transition metals generally form stronger covalent bonds, once formed, any ligand L (including donor atoms of protein side chains other than the proximal histidine) forming a stable Ru–L or Os–L bond will block the labile site.

The stronger bonding to axial ligands, on the other hand, allows the isolation of dinitrogen complexes, e.g. OsN_2(OEP) THF [206] or $Ru(N_2)_2$-(TMP) or RuN_2(TMP)THF [204]. The earlier allusion to a nitrogen carrying osmoglobin seems unrealistic now in view of the abovementioned difficulties encountered in the chemistry of these systems and the fact, that dinitrogen ligands are very labile trans to nitrogen donors such as the imidazole nitrogen of the proximal histidine (see reactions – h, f in Scheme 1).

Polymeric Ru(II) or Os(II) porphyrins with bridging ligands – Substitution reactions with Ru(II) and Os(II) tetrapyrroles extend studies on polymeric iron bridged systems $[M(P)L \wedge L]_x$ of type **H** (M = Fe; L ∧ L means a bidentate donor ligand with two lone electron pairs in opposite directions) which serve to look for organic conductors. A large variety of noble metal porphyrins or phthalocyanines with pyrazine or diisocyanobenzene as bridging ligands L ∧ L is documented in a review from Hanack's group [110]. A recent paper describes the resonant nonlinear optical properties of spin-cast films of the soluble oligomeric diisocyanobenzene-bridged tetrakis(*tert*-butyl)-phthalocyaninatoruthenium(II) complex [216].

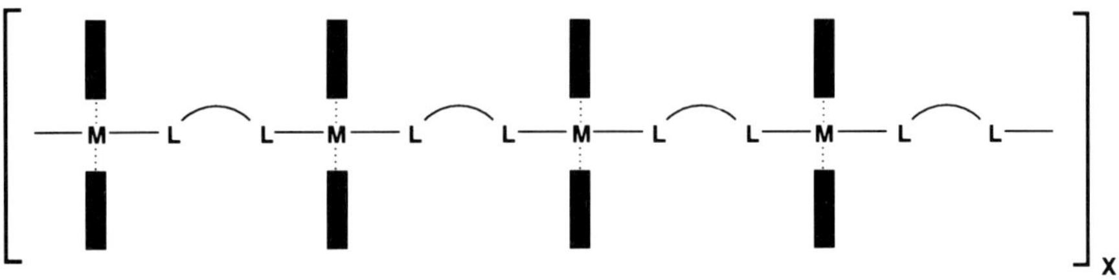

H

Fig. 4. Coordination type **H**, polymeric metal bridged systems $[M\ (L \wedge L)(P)]_x$

An interesting new bridged complex, $[Ru(TDBOHPP)\ L \wedge L]_x$ (type **H**) in which the bridge L ∧ L is 4,4′-azopyridine, has been studied in the search for molecular switches [217]. Protonation of the polymer induces partial oxidation of the Ru(II) to Ru(III) at the expense of the azo groups which are reduced to hydrazo species. Along with the formation of the Ru(II)–Ru(III) mixed valence compound a NIR intervalence band is "switched on". The chemistry of these complexes is further complicated by the phenolic hydroxy groups in the porphyrin ligand which can also be deprotonated and oxidized.

Kinetic investigations – *cis*- and *trans* effects already discussed from a static point of view [190] are also reflected in kinetic measurements. Reaction (11) was studied for combinations of L and L′ = 1-Meim, Py, PBu_3 and $P(OBu)_3$ [218].

$$Ru[P(OBu)_3](Pc)L + L' \rightarrow Ru[P(OBu)_3](Pc)L' + L \qquad (11)$$

The *trans* effect for both FeL(Pc)L′ and RuL(Pc)L′ follows the order $P(OBu)_3 > PBu_3 >$ Py, 1-Meim. The latter ligand has a notable ability to inactivate *trans* ligands. A comparative study of reaction (12) for (P) = (OEP),

(TPP), and (Pc) [219] and L = *t*BuPy or 1-Meim revealed the relative lability of $PhCH_2NC$ complexes to be Ru(Pc) < Ru(P) ≪ Fe(Pc) < Fe(P).

$$Ru(PhCH_2NC)(P)L + BzNC \rightarrow Ru(PhCH_2NC)_2(P) + L \quad (12)$$

A dicyanoruthenium(II) complex is formed pseudo first order kinetics in excess KCN in DMF according reaction (13) [220].

$$Ru(CO)(Pc)DMF + 2CN^- \rightarrow [Ru(CN)_2(Pc)]^{2-} + CO + DMF \quad (13)$$

Bare, tetracoordinate Ru(II) or Os(II) porphyrins and their dimers – An obvious experiment to be done with Ru(II) or Os(II) porphyrins of the type MCO(P) L or $ML_2(P)$ is deligation to obtain the tetracoordinate, "bare" species M(P). This is difficult with CO complexes, the CO ligand being very firmly held (contrary to the hemes). Photolysis as done by Whitten et al. [32] (paths e, f; see Scheme 1) requires an excess of a strong donor ligand to yield bispyridine ruthenochromes [M = Ru, P = OEP, Etio] which can be deligated in vacuo at 220 °C (paths − f, − j, k). Thus, bare Ru(II) derivatives of sterically unencumbered porphyrins do not exist, but dimerize in statu nascendi to form species with Ru–Ru double bonds. The definitive characterization of these dimers required extreme skill in manipulating air-sensitive compounds. This was achieved later by Collman et al. [49, 146] using glove-box techniques. The resulting compounds, $[Ru(P)]_2$ (P = TPP, TTP, OEP) [146] and $[Os(OEP)]_2$ [49] are well-defined; $[Ru(OEP)]_2$ is crystallographically secured (see Table 2), its Ru–Ru distance of 240.8 pm proves a metal–metal bond with a configuration $\sigma^2\pi^4\delta^{nb4}\pi^{*2}$ and $\mu_{eff} = 2.8$ B.M. These dimers are furthermore characterized by detailed NMR investigations.

The $[Ru(P)]_2$ systems are extremely air-sensitive (see Sect. 3.22). The first sample alleged to be $[Ru(OEP)]_2$ [32] was probably one of its autoxidation products, $[RuOH(OEP)]_2O$ (see Sect. 3.22). A molecular thin film claimed to contain porphyrin dimers prepared from RuCO(Meso-IX-DME)Py by irradiation [221] showed UV/Vis spectra similar to those described for the first such product [32]. Therefore, it probably contained also a μ-oxobis(ruthenium(IV)) complex.

The dimers have turned out to be a very interesting and useful class of metalloporphyrins which was extended to dimers containing two divalent metal ions with Rh–Rh single (see Sect. 3.3), Re–Re triple, and Mo–Mo or W–W quadruple bonds [49, 222–224]. These species react with many donor and acceptor molecules (see the following Sects. 3.22 and 3.3). The reactions with typical neutral donor ligands may be seen in Scheme 1, the oxidation reactions in Scheme 2.

Collman has further enriched this chemistry by using cofacial diporphyrins, especially the derivatives of $H_2(DPB)$, 1,8-bis(2,8,13,17-tetraethyl-3,7,12,18- tetramethylporphyrin-5-yl)biphenylene, which contains two etioporphyrin units held together by a biphenylene bridge at two *meso* positions, and has recently reviewed this field [225]. Therefore, the reader is referred to this review for

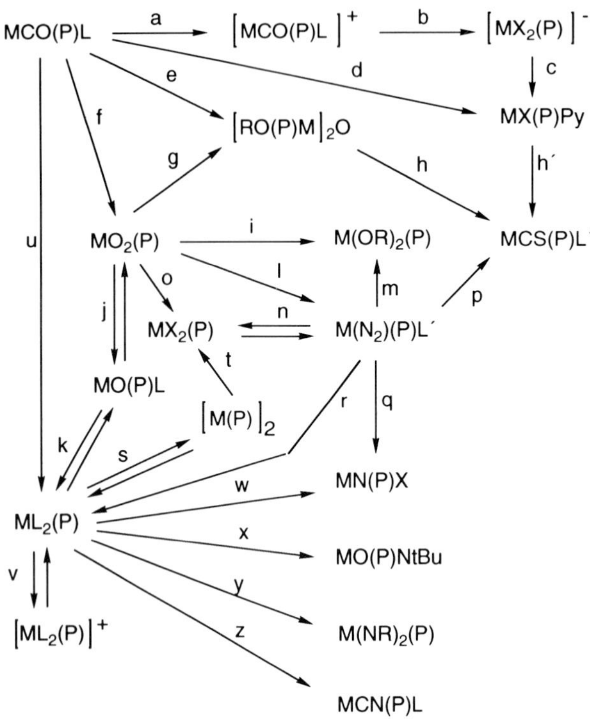

Scheme 2. Reaction paths a – z of ruthenium and osmium porphyrins starting from carbonylmetal (II) complexes MCO (P) L′ (M = Ru, Os). For reaction conditions and references, see Table 5

information on bimetallic ruthenium, osmium, or rhodium complexes designed as catalysts for the catalytic reduction of small molecules like dinitrogen, dioxygen, or protons. A prerequisite is the study of the coordination chemistry of these complexes including the metal–metal interactions in these systems which is described in several recent papers. The binding of dinitrogen, diimine, hydrazine, and ammonia [226–228] and dihydrogen [229–231] to the internally coordinatively unsaturated species, Ru_2L_2(DPB) (L = 1-*tert*-butyl-5-phenyl-imidazole) was investigated.

Asymmetrical, Mixed Bisporphyrinates – The synthetic paths (paths – f, – j, k) preclude an easy synthesis of heterobimetallic complexes like $RuOs(OEP)_2$. Such complexes were obtained in pure forms by stepwise metallation of H_2(DPB) to RuOs(DPB) [232]. Separation of a mixture of $[Ru(OEP)]_2$, $[Ru(OETAP)]_2$, and Ru_2(OETAP)(OEP) was achieved by stepwise oxidation of these bis-metalloporphyrins using $AgBF_4$ in toluene when first $[Ru(OEP)]_2BF_4$ and then $[Ru_2(OETAP)(OEP)]BF_4$ precipitated. The latter was then reduced with cobaltocene to give pure Ru_2(OETAP)(OEP) [233].

Table 5. Details for the redox and successive substitution reactions a–z of Ru and Os porphyrins shown in Scheme 2

Path	M = Ru reactant	Ref.	M = Os reatant	Ref.
a	Anode	[34, 234 235, 236]	Anode	[234]
a	Br_2	[34, 237]		
b	Br_2, X = cN	[238]		
c	HCl, L = Py, X = CN	[238]		
d			Br_2, L = PBu_3 X = Br^-	[239]
e	ROOH	[240]	2,3-DMI/O_2	[161, 240, 241, 242]
f	RCO_3H	[243]	H_2O_2	[244]
	O_2, P = TMP	[245]	RCO_3H	[104, 47]
			tBuOOH	[50]
g	Olefins, R = H	[246]		
h	NaHg, $CSCl_2$, L′ = EtOH	[247]		
h′			NaHg, $CSCl_2$, L′ = Py	[207]
i			$SnCl_2$/MeOH	[244]
			N_2H_4/MeOH	[196]
j	Norbornene, L = EtOH	[246]	P = OEP, PPh_3,	[50]
j, k	H_2N (*t*Bu)	[248]	L = $OPPh_3$	[50]
			P = TPP, PPh_3, L = PPh_3	
j, k	$4PPh_3$, L = PPh_3	[245]		
− k, − j	O_2, L = MeCN	[245]		
l			N_2H_4/H_2O/THF	[47, 195, 206]
m			O_2/L = MeOH; X = OMe	[47, 195, 196]
n			CBr_4, L′ = THF, X = Br^-	[207]
− n			NaHg, N_2	[207]
o	$HNPh_2$, X = NPh_2	[249]	$SOCl_2$, X = Cl	[52, 250a, b]
			X = RO^-, PhS^- or Br^-	[250c]
p			CS_2, L = CS	[207]
q			O_2/MeOH/NH_3, X = OMe	[195]
r			L = NH_3, Py, 1-Meim, THT	[190, 191, 195, 202, 203, 206]
			L = H_2N(*t*Bu)	[248]
t	HBr or HCl,X = Br, Cl	[251]		
u	L = dpm or R_3P	[29, 151]	h*ν*, L = Py	[43, 49, 197]
u			RCl/h*ν*; L,L′ = $P(OMe)_3$	[252, 253]
v	L = L′ = $P(nBu)_3$	[34]	Anode	[190, 202]
	Br_2	[149]	O_2/HX, L = L′	[210]
	O_2	[254]		
w			O_2/MeOH, L = NH_3	[195]
x	L = H_2N(*t*Bu), Br_2	[248]	L = H_2N(*t*Bu), O_2	[248]
			L = $H_2N(p-C_6H_4F)$	[255]
y			L = H_2N(*t*Bu), O_2	[248]
z			O_2, L = $P(OMe)_3$, X = CN	[202]

A true bare ruthenium porphyrin – The sterically encumbered meso-tetramesitylporphyrinato ligand in Ru(TMP) allowed to stop path k (Scheme 1) during the vacuum thermolysis (225 °C; paths – f, j) of $Ru(MeCN)_2(TMP)$ which in turn had been obtained by photolysis of RuCO(TMP) (MeCN) in acetonitrile (paths g, f) [204]. Ru(TMP) gave a variety of axial ligand additions, most notable are the dinitrogen complexes, $Ru(N_2)_2(TMP)$ and $Ru(N_2)(TMP)$ THF; the latter is crystallographically secured (see Table 2). Its NMR spectrum in benzene solution indicated axial asymmetry of the compound; this finding was explained by assuming benzene coordination to the Ru(II) site. Ru(TMP) adds two molecules of L (paths j, f; L = Et_2O, THF). $Ru(THF)_2(TMP)$ was prepared likewise by photolysis from RuCO(TMP)THF and ligated with dinitrogen to yield $Ru(N_2)(TMP)$ THF (paths g, h) [198].

3.2.2 Oxidation and Reduction of Ru(II) and Os(II) Species

The porphyrin complexes of ruthenium and osmium display a rich oxidation–reduction chemistry. Oxidation states + 2, + 3 + 4, and + 6 are well documented. The scope of states that can be realised at the metal is restricted by the fact that the tetrapyrrole ligands $(P)^{2-}$ themselves can be oxidized or reduced to radicals $(P\bullet)^{-1}$ or $(P\bullet)^{-3}$, respectively, at potentials about + 0.7 or – 2.0 V.

Metal(III) species – The synthetic chemistry starts with the monocarbonyls MCO(P)L listed in Table 3. Scheme 2 together with Table 5 survey species that can be obtained from metal(II) species by oxidations or reductions, respectively. The firmly bound carbon monoxide ligands can be removed by diminishing metal-to-carbon π-backbonding. This is initiated by oxidation. Formation of the carbonyl cation (Scheme 2, path a) does not suffice to release CO. The electron is removed from the π-system for M = Ru, producing a π-cation radical, while for M = Os, an Os(III) species is formed. Strong π-acceptor ligands can replace the CO (paths b, c), yielding anionic or neutral metal(III) derivatives. The neutral ones may also directly formed (path d). These oxidations are achieved electrochemically or with bromine.

Metal(IV) species – Oxidation of the carbonyls to metal(IV) derivatives requires oxidants capable of transferring oxygen atoms. Best characterized are the μ-oxobis[oxometal(IV)] complexes directly accessible with alkylhydroperoxides or dioxygen/2,3-dimethylindole (path e). Other metal(IV) derivatives, e.g. bis-alkoxides or oxometal(IV) complexes, are formed from the dioxometal(VI) porphyrins mentioned below.

Metal(VI) species – Hydrogen peroxide or its derivatives (peracids, alkylhydroperoxides) served to obtain the first metal(VI) porphyrin, $OsO_2(OEP)$. Later, $RuO_2(OEP)$, or $RuO_2(TMP)$ which is a very interesting catalyst for the epoxidation of olefins with molecular oxygen (see Sect. 5.2), were made from the carbonyl by action of *m*-chloroperbenzoic acid (path f, R = 3-ClC_6H_4). The water-soluble anion, $[OsO_2(TPPS_4)]^{4-}$, was obtained by treatment of the corresponding carbonyl with trimethylamine-N-oxide [51].

The diamagnetic *trans*-dioxometal(VI) porphyrins are useful starting materials for a variety of reactions. Mild reductants like $SnCl_2$, N_2H_4 or ascorbic acid [249] gave metal(IV) species; in the presence of alcohols or thiols, the bisalkoxides (path i) or bisthiolates (path o, e.g., X = SPh) were formed, respectively. The former displayed temperature-independent paramagnetism (2.3 to 2.7 B.M.), the latter were found to be diamagnetic.

In the presence of neutral donor ligands L (e.g. L = Py, 1-Meim, THT), and an excess of a reductant, e.g. dithionite or hydrazine, dioxo complexes are transformed into bisligandmetal(II) species (ruthenochromes or osmochromes; paths j, k). The logic intermediate in such a four-electron reduction, an oxometal(IV) porphyrin, was only observed for M = Ru (path j, P = OEP, TMP; R = EtOH) when norbornene served as a mild oxygen acceptor.

Dioxoosmium(VI) porphyrins were used to prepare dinitrogenosmium(II) porphyrins (path l, P = OEP, TTP) via reduction with hydrazine hydrate in THF. The latter were transformed into bisalkoxides (path m) by autoxidation in alcohols, in dibromides (path n), thiocarbonyls (path p), nitridoosmium(VI) alkoxides (path q), or osmochromes (path r), respectively.

Metal(V) or metal(VII) species – While the d^5 species [Ru(III) or Os(III)] are stable, especially as osmichrome salts (see below), the other odd electron systems, d^3 or d^1, are labile or not existent at all. The alkoxides, $Os(OR)_2(P)$, were oxidized anodically or with Ce(IV) to yield unstable, but spectrally well-defined cationic Os(V) species $[Os(OR)_2(P)]^+$, while oxidation of $RuO_2(P)$ (P = OEP, TMP) with phenoxathiinylium hexachloroantimonate gave π-cation radicals in which the porphyrin rings were oxidized [256]. Thus, Ru(VII) probably has an oxidation potential which is too high to exist within a porphyrin ring.

Redox chemistry of ruthenochromes and osmochromes – The symmetrical bisligand metal(II) complexes were already presented in Scheme 1 (paths e, f) as products of the photolysis of carbonylmetal(II) porphyrins in the presence of a donor ligand L. Sometimes, this process (transferred to Scheme 2, path u) is difficult to achieve. The oxidation-reduction-substitution sequences given in Scheme 2 (paths f, l, r or f, j, k) give useful alternatives to the synthesis of the complexes $ML_2(P)$. These, in turn, can be used as precursors of the "bare" metal(II) porphyrins (path s). These coordinatively unsaturated species react with all sorts of reagents. In the presence of solvent molecules acting as weak donors, they would autoxidize (paths − s, − k, − j; for Ru, also g), with HCl or HBr, the corresponding porphyrinatoruthenium(IV) halides are formed (path t).

Autoxidation of rutheno- and osmochromes – In the presence of dioxygen, the bisligand metal(II) porphyrins $ML_2(P)$ are autoxidized slowly. If L is a weak donor, e.g. THF or MeCN, an inner-sphere process is operative, ultimately leading to dioxometal(VI) porphyrins (paths − k, − j). These reactions are similar to those of the "bare" ruthenium(II) or osmium(II) porphyrins (see below).

If L is a strong donor (L = Py, 1-Meim), the osmochromes are substitutionally inert. The inner-sphere attack of dioxygen is precluded. Nevertheless, an outer-sphere autoxidation occurs which is acid-induced. It proceeds more rapidly in the presence of Brønsted acids and stops at the Os(III) level (path v); the stoichiometry of Eqs. (14) and (15) has been proved [210, 257].

$$Os(1\text{-}Meim)_2(OEP) + O_2 \rightleftarrows [Os(1\text{-}Meim)_2(OEP)]^+ + O_2^- \qquad (14)$$

$$2\,O_2^- + 2\,H^+ \rightarrow H_2O_2 + O_2 \qquad (15)$$

In the course of the autoxidation in slightly wet pyridine, there is a small, stationary concentration ($\approx$3% of the present osmochrome) of hyperoxide, O_2^-, due to the equilibrium (14). It was detected by quantitative ESR analysis. The hydrogen peroxide formed by dismutation according (15) was photometrically determined as the peroxotitanyl ion in yields between 18 and 71%. The dioxygen consumption in reactions (14) and (15) was measured by respirometry to be $\approx$0.5 mole of O_2 for 1 mole of osmochrome, in accord with the stoichiometry indicated. The reverse of reaction (14) was performed by adding solutions of KO_2 and dibenzo-18-crown-6 in water-free Py or 1-Meim to solutions of the respective osmichrome salts, $[OsL_2(OEP)]PF_6$ (L = Py, 1-Meim), thus establishing that the equilibrium (14) lies to the left.

This reaction is reminiscent to the autoxidation of cytochrome b_5 and described reductions of cytochromes b and c with hyperoxide. In order to study further analogy of osmochrome systems to the cytochromes, water-soluble osmochromes have been prepared from sulfonated precursors. Due to the acidity of pure water, osmochrome systems cannot be held in water, but the osmichrome salt $Na_3[Os[TPPS_4)(1\text{-}Meim)_2]$ was isolated [51].

The autoxidations of ruthenochromes and osmochromes only stop at the metal(III) state if the axial ligands are tertiary amines. The autoxidation of ammine or primary amine complexes $OsL_2(P)$ produces diamagnetic nitrido-osmium(VI) complexes OsN(P) X (path w; L = NH_3, X = OMe, F), or a mixture of diamagnetic osmium(VI) species with oxo and imido groups, i.e. $OsO_2(P)$, OsO(P)N*t*Bu, and $Os(NtBu)_2(P)$ (paths − k, − j, x, y; L = $tBuNH_2$, P = TPP, $\{MeO\}_{12}$TPP. This can be explained by assuming that the osmichrome salts are deprotonated, yielding neutral ammineosmium(III) amides which in further deprotonations and oxidations arrive at the osmium(VI) state. The bis(alkyl-imido) metal(IV) species are susceptible to slow hydrolysis, yielding ultimately the dioxometal(VI) systems. Bis(*tert*-butylamine) ruthenochromes seem to have required bromine for transformation into a Ru(VI) derivative (path x).

These autoxidations do not seem to have studied with secondary amines as axial ligands, but a diamagnetic bis(disphenylamido) ruthenium(IV) porphyrin has been obtained by simple treatment of RuO_2(3,4,5-MeOTPP) with diphenylamine (path o).

In the presence of cyanide, $Os[P(OMe)_3]_2(OEP)$ is autoxidized with loss of one of the donor ligands to give the cyanoosmium(III) complex, OsCN(OEP)-$P(OMe)_3$ (path z) [202].

Autoxidation of "Bare" ruthenium(II) and osmium(II) porphyrins – A resonance Raman study of the intermediates formed during the reaction of "Ru(TPP) " (which was obtained according Scheme 1, paths – f, – j, – k) in toluene [258] proved the anticipated [205] reaction scheme of the inner-sphere autoxidation, the first step of which is the formation of a μ-peroxobis[porphyrinato-ruthenium(III)] complex which is split into two oxoruthenium (IV) fragments. These species precede the formation of μ-oxobisruthenium(IV) porphyrins (reaction 16) for P = TPP, OEP; for P = TMP, a disproportionation (17) is indicated, the resulting Ru(P) itself is further autoxidized.

$$2\,Ru(P) \xrightarrow{O_2} (P)\,RuOORu(P) \rightarrow 2\,RuO(P) \xrightarrow{H_2O} [RuOH(P)]_2O \qquad (16)$$

$$2\,RuO(P) \rightarrow RuO_2(P) + Ru(P) \qquad (17)$$

Cocondensation of sublimed samples of M(TPP) (M = Ru, Os) with dioxygen on a CsCl window in vacuo and subsequent IR transmission spectroscopy at ~30 K of the films thus produced served to identify μ-peroxobis-[porphyrinatometal(III)] species for both metals. Additionally, for M = Ru, the pentacoordinate end-on adduct $Ru(O_2)(TPP)$ was also identified [259]. Its pyridine adduct, $Ru(O_2)(TPP)Py$, was already observed in the previous study [258].

Photochemically and radiolytically activated oxidations and reductions – A variety of photochemically or radiolytically activated oxidations and reductions of osmium porphyrins were investigated [197]. The only reactions not already covered by Scheme 2 are the radiolytic formation of $OsCl_2(P)$ in aerated dichloromethane and the reduction of $Os(OR)_2(P)$ to $Os(ROH)_2(P)$ in the presence of isopropanol.

Reduction of ruthenium(II) and osmium(II) porphyrins – Since the reductive methylation of OsCO(OEP)Py yielded the porphodimethene derivative $OsCO(OEPMe_2)Py$ (see Sect. 2.2), it seemed that reduction took place at the porphyrin ring [100]. However, starting from the dimers $[M(P)]_2$, species are formed that appear to be reduced at the metal [260, 261] as follows from the action of potassium in THF and subsequent alkylation, e.g. with methyl iodide [M = Ru, P = TTP; sequence (18)]. The dianions $[Ru(P)]^{2-}$ are diamagnetic and are attacked at the metal by a variety of electrophiles. $K_2[Ru(TTP)]$ precipitates from the THF solution as a violet-black powder, $K_2[Os(TTP)]$ is a sparingly soluble green salt [260].

$$[M(P)]_2 \xrightarrow{4\,K} 2\,K_2[M(P)] \xrightarrow[-4\,KI]{2\,MeI} 2\,MMe_2(P) \qquad (18)$$

These dianions on stoichiometric protonation, e.g. with H_2O in THF, give hydrido anions which, in turn, with an excess of a Brønsted acid, or with an equivalent of benzoic acid, are transformed into labile dihydrogen complexes finally loosing dihydrogen according reaction sequence (19a, b, c) (M = Ru, Os; P = OEP, TMP, L = THF] [261, 262].

$$K_2[M(P)] \underset{-K^+}{\overset{H^+}{\longrightarrow}} K[MH(P)L] \underset{-K^+}{\overset{H^+}{\longrightarrow}} ML(P)(H_2) \underset{-H_2}{\overset{L}{\longrightarrow}} ML_2(P) \qquad (19)$$

(a) (b) (c)

The hydrido anions are thermally stable; the axial THF may be replaced by other donors (M = Ru, L = 1-*t*Bu-5-PhIm, Py, PPh_3). If the protonation (19b) is done with benzoic acid at − 78°C [230], the dihydrogen complexes $ML(OEP)(H_2)$ (M = Ru, Os, L = THF, 1-*t*Bu-5-PhIm) can be kept and identified in solution at temperatures below − 30 °C. The reverse of (19b) is realized with $Ru(THF)(OEP)(H_2)$ using KOH in THF, with $Os(THF)(OEP)(H_2)$ with lithium diisopropylamide. Even the reverse of (19c) is indicated by formation of $[RuH(OEP)THF]^-$ from $Ru(THF)_2(OEP)$ and 1 atm H_2 in the presence of a large excess of KOH in THF [262].

$Os(THF)(OEP)(H_2)$ decomposes along (19c) within a day. The η-coordinated dihydrogen molecule was well identified in its HD analogues M(THF)-(OEP)(HD) by determination of the HD coupling constants via NMR spectroscopy.

A diruthenium bisporphyrin, $Ru_2(DBP)(1\text{-}t\text{Bu-5-PhIm})_2$, which does not carry an axial ligand within the diporphyrin site, directly binds dihydrogen in benzene to form a bimetallic bridging dihydrogen complex, $Ru_2(H_2)(DBP)$-$(1\text{-}t\text{Bu-5-PhIm})_2$ [231]. From NMR spectra it is deduced that there is a linear array of the Ru–H–H–Ru system in this compound.

3.3 Rhodium and Iridium Porphyrins

The main reactions of rhodium or iridium porphyrins are depicted in Scheme 3 and compiled in Table 6. This comparative table shows that not in all cases have the analogous situations been studied for rhodium and iridium porphyrins; as a whole, a systematic study of iridium porphyrins has commenced only recently. As already mentioned in Table 4, the main starting materials are the aquachlororhodium(III) or carbonylchloroiridium(III) species, i.e. the inspection of Scheme 3 will start from the compounds MCl(P)L (M = Rh; L = H_2O) or MCl(P)CO (M = Ir). Alternative access to the chemistry of rhodium porphyrin chemistry originates at a bare Rh(II) species Rh(P) which is in equilibrium with its metal–metal bonded dimer, $[Rh(P)]_2$ (paths q, − q; see below).

Axial ligand exchange at metal(III) porphyrins – In rhodium(III) porphyrins, the rather firmly bound chloride can be exchanged with hydroxide or rhodanide (path a). The water molecule can be replaced with pyridine or phosphanes (path b); an excess of ligand on prolonged heating furnishes cationic complexes (path g) which themselves, sometimes photolytically, may be subjected to further ligand exchange (path h). The possibility of preparing ligand-bridged binuclear rhodium porphyrins from bidentate donor ligands like bipyridines has also been explored (path i), but not pursued in view of the possibility of obtaining polymeric ligand-bridged cationic systems using pyrazines.

Aromatic and aliphatic thioles RSH (R = 2-, 3-, or 4-tolyl, 2-hydroxyethyl, 2-ethoxyethyl, 4-chlorphenyl, etc.) in the presence of a base yield anionic dithiolatorhodium(III) porphyrins (path j) which show the so-called "hyperporphyrin spectra" and are susceptible to autoxidation yielding hyperoxide ions. Although the formation of the latter ones is formulated via a nucleophilic exchange of coordinated O_2^- with thiolate, it could well be that an outer-sphere electron transfer between he anionic bis(thiolato) complex and molecular dioxygen initiates the observed formation of disulfides RSSR.

Many papers formulate the starting chlororhodium(III) porphyrin just as RhCl(P), as if a *trans* ligand L in MCl(P)L were easily lost (path c, X = Cl). However, the conditions of preparations point to a predominance of hexacoordinate aqua species, RhCl(P)H_2O. Only in one case the formation of a pentacoordinated rhodium(III) halide, the iodide RhI(TMP), seems well-documented [61], see Sect. 2.1.3. The formation of interesting heterobimetallic porphyrins, e.g. (TPP)RhMn$(CO)_5$, [path c, X = Mn$(CO)_5$] was formulated as starting from RhCl(TPP) [63], but the work referred to [264] clearly stated that hexacoordinate species, namely RhCl(TPP)H_2O, RhCl(TPP)(EtOH), or RhCl(TPP)CO were involved. On the other hand, the heterobimetallic species appear to be pentacoordinate about the rhodium, in accord with many metal–metal-bonded porphyrin complexes [222] (see also below).

Bubbling carbon monoxide through a solution of RhCl(TPP)H_2O (path d) yields the photolabile (path −d) carbonylrhodium(III) complex in which the rather high CO frequency (2100 cm^{-1}) suggested the possibility of a nucleophilic attack to this ligand which was indeed achieved (path f), yielding an ethoxycarbonylrhodium(III) moiety; this reaction could be reversed with hydrogen chloride (path −f).

The carbonylchloroiridium(III) porphyrins can be transformed into a variety of other carbonyl complexes by chloride exchange with acids or salts (path e). Concentrated sodium hydroxide in ethanol appears to destroy the carbonyl ligand in these compounds (path −d, a) in a manner similar to the alkoxide addition to RhCl(TPP) CO (path f); here, this should give a carboxylic acid RhCOOH(P) which is decarboxylated to a hydride RhH(P) according to the typical "base reaction" of metal carbonyls. The hydride may then be autoxidized to the hydroxide.

Hydrido- and alkylmetal(III) porphyrins – Treatment of a suspension of IrCl-(OEP) CO in ethanol with sodium borohydride and aqueous 1 M NaOH [paths

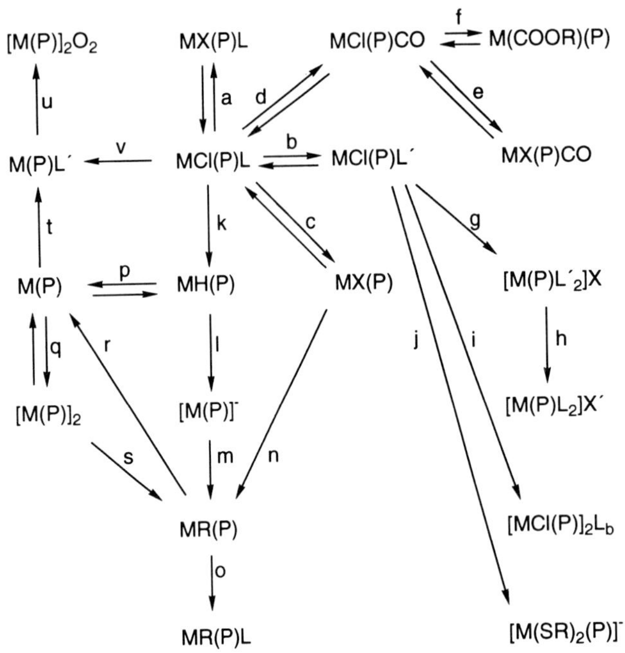

Scheme 3. Reaction paths a–v of rhodium and iridium porphyrins starting from MCl (P) L (M = Rh, Ir). For reaction conditions and references, see Table 6

– d, k (L = EtOH?), l] yielded a red precipitate which after addition of 0.1 M HCl (path – l) gave solid IrH(OEP) [71] as an analytically and spectroscopically well-characterized compound. Similarly, RhH(OEP) was formed [269]; additionally, the very remarkable transformation of RhCl(OEP)H_2O into RhH(OEP) with molecular dihydrogen was stated (path k). Earlier studies on RhCl(TPP) revealed the existence of RhH(TPP) $(H_2O)_2$ [57] which behaves similarly to RhH(OEP).

The stable alkylmetal(III) porphyrins are accessible from the hydrides via their corresponding bases (i.e. Rh(I) or Ir(I) porphyrins) by action of alkyl halides (paths l, m) or from the respective chlorometal(III) porphyrins by action of lithium alkyls (path c, n). They form a very important class of compounds in view of their close relation to the alkylcobalt corrinoids. Of course, the Rh or Ir alkyl group is much more stable than the corresponding Co alkyl group which, in the case of the porphyrins, is photolabile. A variety of axial ligands have been added to the iridium alkyl porphyrins (path o).

Nevertheless, the hydrides or alkyls preferentially crystallize as pentacoordinate species, due to the strong *trans*-effect of the coordinated hydride or alkyl. RhMe(OEP), e.g., has been crystallized from *n*-hexane/dichloromethane solutions in the presence of 1-Meim [172] (path o blocked!).

A notable variant of the formation of the metal methyl bond is the metal insertion to *N*-methyloctaethylporphyrin which proceeds with internal nitro-

Table 6. Details for the reaction paths a–v of Rh and Ir porphyrins depicted in Scheme 3. Negative sign means reverse reaction

Path	M = Rh rectant	Ref.	M = Ir reactant	Ref.
a	$X = OH^-$, SCN^-, $L = H_2O$	[53]		
b	$L' = Py$, $L = H_2O$	[53]		
	$L' = PPh_3$	[60]		
c	$Rh_2(CO)_4Cl_2/I_2$*, X = I	[61]		
	$X = Mn(CO)_5$, $GeCl_3$, $SnCl_3$	[263]		
d	CO	[264]		
– d	hν, $L = H_2O$	[265]		
– d, a			NaOH, $X = OH^-$ $L = H_2O$	[53]
e			H_2SO_4; $X = HSO_4$	[53]
			$AgClO_4$, HBF_4, KCN, KBr	[72]
f	EtOH/EtONa, R = Et	[264]		
– f	HCl	[264]		
g	$L' = PPh_3$(excess), X = Cl	[60]		
g, h	$AgPF_6$; $X' = PF_6$, $L = Me_2NH$	[65]		
h	$L' = Me_2NH$, hν, L = *t*BuNC, PPh_2Me, POMe)$_3$	[58]		
i	$L_b = Py(CH_2)Py$ (n = 0 – 3)	[266]		
	L_b = Bipy	[267]		
j	RSH, $[NBu_4]OH$, P = OEP	[268]		
– d, k, l, – l			$NaBH_4$, HCl, L = CO	[71]
k	$NaBH_4$,HOAc, $L = H_2$	[269]		
	H_2 (via c?)	[269]		
k, l, m			L = CO, $NaBH_4$, NaOH, RI, R = Me, Et, nC_6H_{13}	[71]
c, n			LiMe, X = Cl, L = CO, R = Me	[71]
o			L = CO, Py, NH_3, 1-Meim, R = Me	[72]
			$L = PPh_3$, $R = C_3H_7$	[180]
p, q	Vacuum, 70 °C, 7d	[269]	Anodic oxid. on graphite	[270]
p, q	O_2 in C_6H_6	[269]		
r	hν in C_6H_6,	[271]		
r, q			hν in C_6H_6, R = Me, P = OEP	[272]
s	$PhCH_2Br$, – RhBr(P)	[269]		
s. – q, – p			MePh (neat), 120 °C	[272]
– q, t	NO, L' = NO	[273]		
	$L' = O_2$, – 80 °C	[273]		
t, u	$L' = O_2$, 20 °C	[273]	$L' = O_2$ on graphite	[270]
v	L' = NO	[273]		

* Result of metal insertion

gen-to-metal migration of the methyl group for both Rh and Ir systems [56, 71], yielding RhMe(OEP) or IrMe(OEP).

Metal(II) species, homobimetallic complexes – During a rhodium insertion into H_2(TPP) using $[Rh(CO)_2Cl]_2$ in HOAc/NaOAc, a paramagnetic rhodium(II) porphyrin was observed [57] which could be transformed into a hydridorhodium(III) porphyrin with dihydrogen (path – p) and subsequently into its corresponding base, the anion $[Rh(TPP)]^-$, which may be likewise regarded as a Rh(I) species (path l). The existence of a dimerization equilibrium (path q) was not excluded. Later, the paramagnetic species by its ESR spectrum was identified to be the dioxygen adduct, $RhO_2(TPP)$ [273] which is formed from the dimer, $[Rh(TPP)]_2$ (paths $-$ q, t) and adventitious presence of dioxygen.

A clean formation of $[Rh(OEP)]_2$ proceeds via thermolysis [269] or photolysis [273] with loss of dihydrogen from or autoxidation of the hydride RhH(OEP) (path p). The tetramesitylporphyrin complex, Rh(TMP) [61], does not dimerize at all due to the sterically hindrance created by the two *ortho*-methyl groups of each phenyl ring (see Ru(TMP)!), however, the *meta*-methyl groups of the rhodium(II) derivative prepared from tetra-kis(3, 5-xylyl) porphyrin $[H_2(TXP)]$ do not prevent dimerization, and the complex is isolated as a dimer $[Rh(TXP)]_2$ which dissociates (path $-$ q) prior to chemical reactions. Photolysis of RhMe(TMP) [274] (path r) is another suitable access to Rh(TMP) [271].

The dissociation of $[Rh(TPP)]_2$ was achieved photochemically in 2-methyltetrahydrofuran at 77 K; Rh(TPP) was identified by the characteristic perpendicular component found in the ESR spectrum of the irradiated sample at $g = 2.46$; dimerization seems to be retarded due to the high viscosity of the solvent at 77 K [275a]. It may be noted in this context that the dirhodium porphyrin $[Rh(CO)_2]_2(TPP)$ formed as the primary product in metal insertions (see Sect. 2.1.3) on photolysis in benzene yields the dimer $[Rh(TPP)]_2$ [275b]; for this reason, the latter is a good entrance to Rh(TPP) chemistry as stated before. Photolysis of $[Rh(CO)_2](TPP)$ in pyridine was claimed to give a pyridine adduct $RhPy(TPP) * 2H_2O$ (paths $-$ q, t) [276c], in benzene-carbon tetrachloride, however, illumination causes oxidation, yielding RhCl(TPP)CO [276].

Dimerization was also achieved from Rh(III) porphyrins, just by deligating RhI(OEP) [277] with silver salts like $AgBF_4$ in rigorously dried dichloromethane in the absence of any donor ligands. The bare cation $[Rh(OEP)]^+$ dimerized to a diamagnetic dication $[Rh_2(OEP_2)]^{2+}$. In this bis[porphyrinato-(-1) rhodium(II)] complex two porphyrin π-radical anions $(P^{\bullet})^-$ are held together by a Rh–Rh single bond and π–π-interactions; the latter were evident from the typical visible absorption at $\lambda_{max} = 793$ nm and abnormally high chemical shifts in the ^1H NMR spectrum.

Since cobalt(II) porphyrins react both with nitric oxide and dioxygen, it is not surprising that adduct formation of these small molecules also occurs with rhodium(II) porphyrins (paths $-$ q, t, $L' = NO, O_2$) [273]. The binding appears to be irreversible. The products have been identified by UV/Vis, IR, and ESR spectra and are formulated as Rh(III) derivatives, containing the NO^- or O_2^- ions, respectively, as single axial ligands in pentacoordinate species. $RhO_2(OEP)$ forms base adducts $RhO_2(OEP)L$ [L = Py, Pic, or $P(OR)_3$] at 193 K as is seen from ESR data. Warming up toluene solutions of $RhO_2(OEP)$ to 293 K yields a diamagnetic complex which is formulated as $[Rh(OEP)]_2O_2$.

The coordinated nitroxide (NO^-) is slowly autoxidized to coordinated nitrite. RhNO(OEP) is also formed by action of NO upon the hydride RhH(OEP) (paths p, t) or even RhCl(OEP) (path v) in toluene. The latter reaction proceeds via an initial formation of a nitrosyl chlororhodium(III) porphyrin radical which reacts with a second NO molecule liberating nitrosyl chloride.

While carbon monoxide does not react with cobalt(II) porphyrins, it readily does so with $[Rh(P)]_2$, e.g. in benzene. Most of these subsequent reactions are complicated and do not fit into Scheme 3; they are reported in Sect. 4.2. Just the reactions of the highly sterically hindered monomeric Rh(P) complexes (P = TMP, TTiPP) yield simple monocarbonyls (Scheme 3, path t, L′ = CO) [271]. Because of the 17-electron configuration of Rh(II), these species are remarkable; from the ESR spectra displaying rhombic character the existence of a bent Rh–CO moiety is concluded. These moncarbonyls behave like acyl radicals (see Sect.4.2).

One of the most interesting reactions of Rh(II) porphyrins is the setting up of equilibria like (20) with methane [61] (paths $-$q, and then $-$p and $-$r in parallel; R = Me), yielding rhodium(III) hydride and alkyl moieties at the same

$$[M(P)]_2 \rightleftarrows 2\,M(P) \xrightleftharpoons{CH_3R} MH(P) + M(CH_2R)(P) \qquad (20)$$

M = Rh, Ir; R = H, Ph, COOEt, etc.; P = OEP, TMP, TXP (see also Sect. 4.2)

time. The thermodynamics and kinetics of these reactions were studied; for P = TMP or TXP, the equilibrium constants at 353 K and 1 atm of CH_4 were determined by NMR measurements to be K = 7300 or 0.034, respectively, in the direction of methane fission; for P = TMP at room temperature and 1 atm of CH_4, K was too large to be determined. Benzene solutions of RhH(TMP) and RhMe(TMP) at 353 K give reductive elimination to form Rh(TMP) and CH_4 (paths r and p in parallel). Toluene is also attacked at 398 K in 24 h by $[Rh(OEP)_2]_2$ (paths $-$q, $-$r, $-$p), giving a 50% yield of RhH(OEP) and $RhCH_2Ph(OEP)$ [278]. Thus, contrary to the general opinion that CH bonds of benzyl groups are less stable than those of methane, methane appears to be much more reactive in this system. The reason for this is seen in a concerted attack of two rhodium porphyrin fragments approaching the methane molecule from opposite faces; such a topology would suffer from steric hindrance in phenylmethane.

Aromatic hydrogen atoms are not attacked at all by Rh(II) porphyrins. See, however, electrophilic substitution with Rh(III) porphyrins (Sect. 4.2).

A variety of other organometallic reactions starting from the homodimers are also reported in Sect. 4.2.

In order to prepare a dirhodium compound, $Rh_2(DPB)$, the cofacial bisporphyrin, $H_4(DPB)$, which was introduced in Sect. 3.2.1, was metallated to yield dirhodium(III) compounds with iodide and/or ethoxycarbonyl as axial ligands [279]. While iodide is supposed to be able to reside inside the diporphyrin cavity, the ester moiety is located outside the cavity, since the compound $(RhCOOEt)_2(DPB)$ does not add triethylphosphane, contrary to the diodide. Photolysis of all Rh(III) derivatives obtained yielded the wanted product,

$Rh_2(DPB)$. In contrast to the dimers $[Rh(P)]_2$, this compound reacted with dihydrogen only after intermediate addition of CO to yield a hydride $(RhH)_2(DPB)$ with two Rh–H bonds probably located inside the cavity, An insertion of CO into the Rh–H bonds, as with RhH(P) (see Sect. 4.2), was not observed. The dihydride was able to be dehydrogenated with molecular oxygen to reform $Rh_2(DPB)$; the formation of a dioxygen adduct (μ-peroxo derivative) of the dirhodium(II) species which might be expected according to Scheme 3 (paths − q, t, u) was not observed.

The dimeric iridium(II) porphyrinates, $[Ir(P)]_2$, are far less well studied [222, 270] than their rhodium analogs. The formation of $[Ir(OEP)]_2$ is cleanly achieved by photolysis of IrMe(OEP) (path r, q) [272]. Hydrogenolysis of the dimer (paths − q, − p) yields the hydride, neat toluene a mixture of the hydride and the benzyliridium(III) compound [paths − q, − p, − r, similarly to Rh(II) porphyrins].

Heterobimetallic species – Nucleophilic substitution of InCl(OEP) with $[Rh(OEP)]^-$ gave the heterometallic homoleptic dimer, (OEP)RhIn(OEP), which was confirmed by X-ray crystallography [280] and reviewed [222]. As already cited [263], nucleophilic substitution of RhCl(TPP) could also be achieved with carbonylmetallate anions, thus yielding heterobimetallic heteroleptic complexes, e.g. RhX(TPP) with X = $Mn(CO)_5$, $Co(CO)_4$, $SnCl_3$, or $GeCl_3$, respectively (path c). Similar compounds may be prepared by metathesis; on mixing the homoleptic homometallic complexes $[Rh(OEP)]_2$ and $[Mn(CO)_5]_2$, $[MoCp(CO)_3]_2$, or $[RuCp^*(CO)_2]_2$, heterometallic, heteroleptic compounds Rh(OEP)R with R = $Mn(CO)_5$, $MoCp(CO)_3$, or $RuCp^*(CO)_2$, respectively, were obtained (path s) showing Rh–Mn, Rh–Mo, or Rh–Ru bonds [281].

Possible analogies to heme – Reconstitution of sperm whale apomyoglobin with rhodium(III) complexes of meso- and deuteroporphyrin afforded stable rhodium(III) myoglobins in a 1:1 (Rh to protein) stoichiometry [282]. The starting materials, RhI(Meso-DME) $*$ $2H_2O$ and RhI(Deut-DME)H_2O, were prepared by the Grigg method (see Sect. 2.1.3) [62]. RhMe(Meso-DME) was also prepared. These three rhodium(III) porphyrin dimethylesters were saponified and the corresponding disodium dicarboxylates treated with apomyoglobin at pH 7. The "rhodioglobins" ["Rh(P)G"] were purified by gel filtration. From the reluctance of [Rh(Meso)G] to react with added ligands like F^-, OCN^-, N_3^-, or CN^-, it was concluded that both the proximal and distal histidines act as ligands to Rh(III) in [Rh(Meso)G], thus forming an internal hemichrome-like species. In the methylrhodium(III) derivative, [RhMe(Meso)G], the prosthetic group is only weakly held by the proximal histidine, and the methyl group seems to be compressed. Hence, on incubation of native metmyoglobin with $[Rh(Meso)]^+$ or RhMe(Meso) for 24 h at 35 °C, the former replaced the heme to about 90% while the latter went in to an extent of only 15%. The NMR signals of the RhMe group at − 5.5 and − 7.5 ppm where assigned to the normal and denatured protein, respectively, demonstrating that RhMe(Meso) might be used as a NMR probe to get information about heme sites in proteins.

Molecular recognition with rhodium complexes of functionalized porphyrins – The coordination abilities of rhodium(III) porphyrins outlined in Scheme 3 (paths b and g) were used to design porphyrins with special receptor properties, i.e. *cis*- or *trans*-5, 15-bis(2-hydroxy-1-naphthyl)octaethylporphyrin [H_2(npOEP)], *trans*-5, 15-bis(8-quinolyl) porphyrin, or tetrakis(2-hydroxyphenyl) porphyrin. After Rh insertion, these porphyrins provide lateral OH or N donor groups at a fixed distance from the coordination site at the Rh atom. Unlike RhCl(TPP), RhCl(npOEP) activates acetone in a manner that an α-metallation of acetone takes place, yielding Rh(CH_2COMe)(npOEP) [283].

A two-point fixation of amino acids and amino esters in non-ionic forms via simultaneous metal-coordination and hydrogen bonding interactions was observed with RhCl(npOEP) which in $CDCl_3$ was found to extract 1 mol of free amino acids such as phenylalanine or leucine from water at pH 7 to form adducts irreversibly [284], the amino group forming a stable Rh–N bond.

A reversible formation of adducts with amino acids or their esters occurred with the above-mentioned acetonide, Rh(CH_2COMe) (npOEP) in which one of the naphthyl hydroxy groups forms a hydrogen bond to the acetonyl oxygen atom, the other one a hydrogen bond to the carboxy oxygen atom of the bound amino acid or its ester. Obviously, the strong trans effect of the Rh–C bond in this complex weakens the Rh–N bond of the coordinated amino acid. Rh(CH_2COMe)(npOEP) was used as a carrier to transport relatively lipophilic amino acids, e.g. phenylalanine, leucine, isoleucine and norleucine, through an artificial liquid trichloromethane membrane. The thermodynamic data of these partition equilibria were determined [285].

Replacement of the 2-naphthyl groups by 2-dimethylaminomethylphenyl groups in H_2(npOEP) also led to a rhodium porphyrin being able to extract leucine from water, however, the situation is complicated by dimerization of the rhodium porphyrin due to intermolecular amine–rhodium bonding [286]. A rhodium complex of a trifunctional chiral bis(2-hydroxynaphthyl)porphyrin related to the above-mentioned RhCl(npOEP) system was used to separate diastereomers formed via two-point fixation of amino acids [287].

The effect of the lateral OH groups in Rh(III) porphyrins was enhanced by silver ion coordination to these OH groups. Thus, the methanolysis of *p*-nitroacetanilide or ethyl acetate was accelerated in the presence of *trans*-RhCl-(npOEP) and $AgClO_4$ [288]. A study of binding of nucleobases to various rhodium(III) porphyrins was also done but did not disclose any enhanced specificity which might have been caused by lateral functional groups [289].

Rhodium phthalocyanines – In view of the current interest in polymeric bridged phthalocyanines [110], Rh(III) phthalocyanines with trans-bidentate ligands were synthesized. RhCl(Pc)L (obtained from *o*-cyanobenzamide) [290] appears to strongly hold its chloride ligand. Action of ligands L′, namely pyridine,

4,4′-bipyridine, 2-methylpyrazine, and 1,4-diazabicyclooctane yields products according to Scheme 3 (path b) [291]. The formation of anionic rhodium(III) complexes exemplified for porphyrins in Scheme 3 (path j) finds its phthalocyanine counterpart in the formation of K $[Rh(CN)_2(Pc)] * 2H_2O$ from RhCl(Pc)L and KCN in acetone [290]; it is amazing that this reaction appears to be unknown for porphyrins. Simple boiling of the dicyanorhodium(III) salt with water extracts KCN and leaves the polymeric μ-cyano(phthalocyaninato)-rhodium(III) $[RhCN(Pc)]_x$ as a blue powder which shows a dark conductivity of $\sigma = 4 \times 10^{-4}$ S/cm and semiconducting behaviour. This material can be depolymerized by treatment with pyridine for 3 d at 100 °C, yielding RhCN(Pc)Py.

3.4 Palladium, Platinum, Silver, and Gold Porphyrins

Pd(II), Pt(II), and Au(III) ions which strongly prefer square planar coordination geometry in their d^8 configuration, but also the Ag(II) ion in its d^9 configuration, are practically devoid of any coordination chemistry in their porphyrin complexes, apart from intermediates that may be observed during metalloporphyrin formation and a few reactions which have already been discussed in previous reviews [7, 8].

The only reactions that do occur with these metalloporphyrins follow oxidations of the metals or the porphyrin in rings. They have been known for a long time and summarized in reviews on the primary redox processes of metalloporphyrins [292, 293] or electrochemistry in nonaqueous media [294].

The formation of dichloro(porphyrinato)platinum(IV) porphyrins was achieved by action of hydrogen peroxide in hydrochloric acid (Eq. 21). Apart from reduction to Pt(II), further reactions of the chlorides have not been observed.

$$\mathrm{Pt(P)} \xrightarrow{\mathrm{H_2O_2/2HCl}} \mathrm{PtCl_2(P)} \quad (\mathrm{P = OEP,\ TTP\ [85]}) \tag{21}$$

The analogous reactions of Pd(OEP) or Pd(TTP) did not result in formation of Pd(IV) porphyrins, but in excessive chlorination of the *meso* [295] or peripheral (β-pyrrole) positions [296] of the porphyrin rings. The oxidation of Pt(P) in the absence of hydrogen chloride did not lead to isolatable products. Of course, an oxoplatinum(IV) porphyrin would have been a very interesting compound.

Due to poor chemistry and the stability of the porphyrin complexes containing the precious metals Ag, Au, Pd, and Pt, most papers dealing with these complexes are devoted to physical measurements, especially optical or photophysical investigations [23].

The metal-to-porphyrin backbonding [190] increases in the series Ni(II) < Pd(II) < Pt(II) for the corresponding metalloporphyrins. It was speculated that in the same sense, due to the π-donation to the porphyrin ring, electrophilic substitution might be accelerated. However, a systematic study of the relative reaction times of the Vilsmeier formylation of a variety of tetra-*p*-tolylporphyrin complexes revealed that the total charge density of the porphyrin ligand in its metal complex is the determining factor, i.e. the more polar the metal–nitrogen bond, the more rapidly electrophilic substitution occurs [77].

The comparative photophysics of Pt(II) and Pt(IV) porphyrins [297] have been investigated. A correlation of X-ray crystallographic data and spectroscopic properties was done for a variety of octaethylporphyrin complexes of divalent metals, incorporating those of Ni, Pd, Pt, Cu, and Ag, giving useful UV/Vis and IR data for the synthetic chemist as well [298].

Silver porphyrins are interesting because of the existence of disilver(I), paramagnetic silver(II), and diamagnetic silver(III) complexes which has been known for a long time [7, 8]. The latter are favorably prepared by electrochemical [299] or peroxodisulfate [300–302] oxidation of silver(II) porphyrins. The Ag(III) porphyrins were used for ^{1}H and ^{2}H resonance experiments [299] on one hand and for the determination of monomer–dimer equilibria [301] or self-exchange rate constants [301] at water-soluble silver porphyrins.

Treatment of red, ionic [Au(TPP)]Cl with aqueous sodium hydroxide in DMSO yielded a greenish brown product which was isolated by extraction with *n*-hexane. It was identified as a product of the nucleophilic addition of hydroxide to a *meso* (5-) position of the TPP ring, yielding a 5-hydroxy-5-*H*-porphyrin (hydroxyphlorin) derivative, Au(HOTPP) [303].

An interesting application of the square planar metalloporphyrins or phthalocyanines is their incorporation into coplanar stacks which on partial oxidation become electrically conductive materials [24, 25, 110]. The palladium(II) complex of *meso*-tetramethylporphyrin, Pd(TMeP), was incorporated into a one-dimensional stack by electrochemical oxidation in the presence of $NBu_4[ReO_4]$ (electrocrystallization), yielding the semiconducting π-radical cation salt [Pd(TMeP)] $[ReO_4]$ [304]. Gold(III) porphyrin salts of the type [Au(TPP)] $[M(mnt)_2)]$ [M = Ni, Pt, Au; $(mnt)^-$ = maleonitrile dithiolate ion] were obtained by deposition from DMF/ MeCN mixtures. From the known structure of the related compound, $[Au(TPP)]_2[Pt(mnt)_2]$ and from ESR and magnetic data, it was concluded that the new semiconducting compounds consisted of alternate stacks of the constituent metal chelate cations and anions [305].

4 Organometallic Reactions of Noble Metal Porphyrins

In this section, some organometallic reactions will be presented which could not be incorporated into the inorganic reactions and which are also important for catalytic reactions. Most of the contributions have come from the schools of Collman, Guilard, Ogoshi and Wayland; many of them have been reviewed already [222, 223, 225, 306]. Not only simple alkyl or aryl derivatives, but olefin, carbene, and silylene complexes have recently been documented.

4.1 *Organometallic Reactions of Ruthenium and Osmium Porphyrins*

Alkyl and aryl systems – As already mentioned in Sect. 3.2, dialkyl ruthenium and osmium porphyrins have been synthesized according to Eq. (22) by the reaction of metalloporphyrin dianions $[M(P)]^{2-}$ with alkyl iodides [223, 260, 261, 307]. These dianions have been obtained by reduction of porphyrin dimers $[M(P)]_2$.

$$[M(P)]^{2-} + 2\ RX \rightarrow MR_2(P) + 2X^- \tag{22}$$

$$M = Ru, Os; \quad RX = CH_3I, C_2H_5I; \quad P = OEP, TTP.$$

The resulting dialkylmetal(IV) derivatives are thought to be diamagnetic. The diethylruthenium complex $Ru(C_2H_5)_2(P)$ was thermally labile at room temperature, and a mixture of the ethylidene species $Ru(CHCH_3)$ (P) and the Ru(III) compound $RuC_2H_5(P)$ was formed via independent radical processes [307] (Eq. 23).

$$3\ Ru(C_2H_5)_2(P) \rightarrow Ru(CHCH_3)(P) + 2\ RuC_2H_5(P) + C_2H_4 + 2\ C_2H_6 \tag{23}$$

Some diarylruthenium(IV) species were made from $RuBr_2(TPP)$ or $RuBr_2(OEP)$ with LiPh [157, 251] (Eq. 24).

$$Ru(P)Br_2 + 2\ LiPh \rightarrow RuPh_2(P) + 2\ LiBr \tag{24}$$

Thermolysis of $RuPh_2(TPP)$ at 100 °C for 30 h in benzene produces a phenylruthenium(III) complex, RuPh(TPP) [155] (see Table 2). In contrast to $RuPh_2(P)$ and FePh(P) [222] no *N*-arylation occurred on oxidation of the five-coordinate RuPh(P) [306].

Oxidation of $RuAr_2(OEP)$ (Ar = p-XC_6H_4 where X = H, Me, OMe, Cl, F) with $AgBF_4$ in toluene led to the formation of the corresponding N-aryl porphyrin complex salts [RuAr(OEPNPh)] BF_4 [153, 308, 309]. The oxidation of the dimethylruthenium compound $RuMe_2(OEP)$ also gave a rather labile cationic *N*-alkyl- porphyrin complex, $[RuMe(OEPNMe)]^+$, which was slowly converted to a methylruthenium porphyrin cation $[RuMe(OEP\text{-}N\text{-}\mu\text{-}CH_2\text{-})]^+$ in which a methylene bridge was sitting between the Ru atom and one of the N atoms of the OEP ligand [310].

Reduction of $RuMe_2(OEP)$ with sodium naphthalenide, NaNp, led to decomposition of the anion thus formed (sequence 25) [308].

$$RuMe_2(OEP) \xrightarrow{NaNp} [RuMe_2(OEP)]^- \xrightarrow{-CH_3} [RuMe(OEP)THF]^- \xrightarrow{O_2} RuMe(OEP) \tag{25}$$

The anionic methylruthenium(II) species was autoxidized to a Ru(III) compound, RuMe(OEP). The methyl group of this compound was accidentally transformed into a coordinated carbon monoxide molecule by an excess of 2,2,6,6-tetramethylpiperidine-1-oxyl (TEMPO) [158] on an attempt to use TEMPO as a radical trap for the measurement of the Ru–C bond energy in solution. This was the first transformation of a methyl group to carbon monoxide to be observed in the proximity of a metal.

Reaction of neopentyllithium, Li(neoPe), with $RuCl_2(OEP)$ yielded paramagnetic Ru(neoPe)(OEP) which was treated with further Li(neoPe) to give $[Ru(neoPe)(OEP)]_2(\mu\text{-}Li)_2$, in which the two Li atoms are sandwiched between two metalloporphyrin moieties in a very remarkable structure [311].

Carbene and silylene complexes – Carbene complexes of ruthenium [312] and osmium [313] porphyrins were formed from the neutral dimers $[M(P)]_2$ with diazoalkanes (Eq. 26).

$$[M(P)]_2 + 2\ N_2CRR' \rightarrow 2M(P)\ CRR' + 2\ N_2 \tag{26}$$

M = Ru, Os; P = OEP, TTP; R, R′ = *p*-Tolyl; R = H, R′ = $SiMe_3$, CO_2Et.

Instead of adding 4-substituted pyridine ligands *trans* to the carbene ligand, the complexes Os(CHR′)(TTP) (R = $SiMe_3$, CO_2Et) add these donors to the carbene carbon atom with formation of an ylide species. This reaction is not observed for $Os(CR_2)(TTP)$ with R = Ph [313c]. Nevertheless, the diphenylcarbene or trimethylsilylmethylene osmium porphyrins crystallize as Os(CRR′)-(TTP)THF with a THF molecule in *trans* position.

Ruthenium carbene complexes were also formed from the reaction of gem-dihalides R_2CCl_2 with the dianion $[Ru(P)]^{2-}$ (Eq. 27) [261] and by insertion of ruthenium into a metal-free porphyrin containing an *N,N′*(2,2-diarylvinylidene) bridge [314, 315]. Along with the formation of this metal diarylvinylidene ruthenium porphyrin, two rutheniumdicarbonyl complexes of porphyrins were formed in which the *N,N′* bridge remained intact, but the ruthenium had inserted into a pyrrole C–N bond. These damaged porphyrins could be reverted to vinylidene complexes on heating.

$$[Ru(P)]^{2-} + R_2CCl_2 \rightarrow Ru(CR_2)\,(P) + 2\ Cl^- \tag{27}$$

Base stabilized silylene complexes, $Os(SiR_2THF)(TTP)$ (R = Me, Et, *i*-Pr) [167], have been prepared from $[Os(TTP)]_2$ and hexamethylsilacyclopropane or K_2 [Os(TTP)] and dialkyldichlorosilanes. The X-ray structure determination (see Table 2) of a THF adduct, $Os(SiEt_2THF)(TTP)THF$ clearly showed the silicon- and osmium bound THF molecules. NMR analysis of a solution of $Os(SiR_2THF)(TTP)$ in C_6D_6 showed that a stoichiometric amount of pyridine replaces the silicon-bound THF; however, on concentration of such solutions, partial displacement of the silylene ligand by pyridine occurs, and taking to dryness produces a 7:3 mixture of $Os(SiR_2Py)(TTP)$ and $OsPy_2(TTP)$.

Olefin complexes – Treatment of $[Ru(P)]_2$ with ethylene yielded the π-complex $Ru(CH_2{=}CH_2)\,(P)$ [260] which was also obtained by the reaction of $[Ru(P)]^{2-}$

with 1,2-dibromoethane [261]. Very remarkably, $Ru(CH_2{=}CH_2)(P)$ was formed likewise when $[Ru(P)]^{2-}$ was reacted with dichloromethane in THF or $[Ru(P)]_2$ with diazomethane [260]. Since, according to Eqs. (25) or (26), these reactions ought to have led to a carbene complex with a methyleneruthenium unit, it appears that this entity is unstable and exchanges one methylene fragment, yielding the ethylene complex $Ru(CH_2{=}CH_2)(P)$ and a weakly coordinated Ru(P) unit, $Ru(P)(THF)_2$.

Sterically hindered systems – As already pointed out in Sect. 3.2, the sterically hindered ruthenium tetramesitylporphyrins Ru(TMP) or $Ru(TMP)L_2$ (L = MeCN, N_2, THF) [204, 154] are useful starting materials for the synthesis of organometallic compounds [209, 316]. While the reaction of $Ru(N_2)_2(TMP)$ with acetylene yielded a divinylidene bridged complex, (TMP)Ru=CH–CH=Ru-(TMP), 1:1 π-complexes were formed with phenylaacetylene, diphenylacetylene, ethylene, or cyclohexene [209]. $Ru(CH_2{=}CH_2)(TMP)$ was deposited solvent-free by evaporation from benezene and precipitated with 2-propanol as a diliganded solvate, $Ru(CH_2{=}CH_2)(TMP)$ *i*-PrOH ∗ *i*-PrOH. The compounds were characterized by elemental analyses and ^1H and ^{13}C NMR spectra.

The reaction of Ru(TMP) with ethyl diazoacetate yielded a carbene complex, e.g. $Ru(CHCO_2Et)(TMP)$ [316]. An excess of the diazo compound led to catalytic formation of *cis*- and *trans* diethyl maleate in an unexpected ratio of 15:1. The nucleophilic ethyl diazoacetate is proposed to attack the electrophilic carbene complex and produce an intermediate betaine-line species which eliminates both Ru(TMP) and N_2 to form the maleates. Similar reactions were observed with Os(TTP) complexes [313a]. These reactions are reminiscent of the above-mentioned lability of a putative methyleneruthenium porphyrin.

4.2 Organometallic Reactions of Rhodium and Iridium Porphyrins

Alkylmetal(III) porphyrins – The formation of the typical alkylrhodium(III) or -iridium(III) species has already been discussed in Sect. 3.3, Scheme 3 (paths c + n, k + 1 + m, − r, s) and Table 6.

Oxidative addition of RX to bis[dicarbonylrhodium(I)] porphyrins [317] (see Eq. 28) or monorhodium(I) porphyrins (Scheme 3, path m) also produces σ-bonded complexes. The organic substrates RX include aldehydes, anhydrides, aryl or acyl, arylcarbonyl, or ethoxycarbonylmethyl halides, cyclopropyl ketones [62] or highly strained cyclopropanes [318]. The fate of the second rhodium ion formulated as $[Rh(CO)_2]^+$ in Eq. (28) was not investigated.

$$[Rh(CO)_2]_2(P) + RX \rightarrow RhR(P) + [Rh(CO)_2]^+ + X^- + 2\,CO \qquad (28)$$

Acetylrhodium(III) tetraphenylporphyrin was formed on warming *N*,*N*-dimethylacetamide solutions of RhH(TPP) [319]. Obviously, an acetyl group had been abstracted from a solvent molecule.

Reactions of rhodium porphyrins with diazo esters – According to Callot et al., iodorhodium(III) porphyrins are efficient catalysts for the cyclopropanation of alkenes by diazo esters [320, 321]. The transfer of ethoxycarbonylcarbene to a variety of olefins was found to proceed with a large *syn*-selectivity as compared with other catalysts. In their study to further develop this reaction to a shape-selective and asymmetric process [322], Kodadek et al. [323] have delineated the reaction sequences (29, 30) and identified as the active catalyst the iodoalkylrhodium(III) complex resulting from attack of a metal carbene moiety Rh(CHCOOEt) by iodide.

$$\mathrm{RhI(TTP)} \xrightarrow{\mathrm{N_2CHCOOEt}} [\mathrm{Rh(CHN_2COOEt)(TPP)}]^+\mathrm{I}^- \xrightarrow{-\mathrm{N_2}} [\mathrm{Rh(CHCOOEt)(TTP)}]^+\mathrm{I}^- \qquad (29)$$

$$[\mathrm{Rh(CHCOOEt)(TTP)}]^+ + \mathrm{I}^- \rightarrow [\mathrm{Rh(CHI\{COOEt\})(TTP)} \qquad (30)$$

The product of reaction (30) is thought to coordinate a further molecule of ethyldiazoacetate trans to the iodoalkyl group which looses dinitrogen, yielding a hexacoordinate rhodium carbene complex according to Eq. (31) which transfers its carbene moiety to an attacking alkene molecule.

$$\mathrm{Rh(CHI\{COOEt\})\,(TTP)} \xrightarrow{\mathrm{N_2CHCOOEt}} \mathrm{Rh(CHI\{COOEt\})(CHN_2COOEt)\,(TPP)} \xrightarrow{-\mathrm{N_2}} \mathrm{Rh(CHCOOEt)(CHI\{COOEt\})(TPP)} \qquad (31)$$

Although σ-alkylrhodium(III) porphyrins do not very readily add further ligands (see Scheme 3, path o) due to the strong *trans*-labilizing effect of σ-alkyl groups, the σ-donor strength of the ethoxycarbonyliodoalkyl group is certainly reduced due to the presence of the iodo and ethoxycarbonyl substituents. The propensity of the empty *trans* coordination site of the iodoalkylrhodium(III) species towards binding of an organometallic donor is demonstrated by its NMR spectrum in the presence of excess styrene which clearly shows than an adduct Rh(CHI{COOEt}) (TPP) ($CH_2{=}CHPh$) is formed [323b]. Furthermore, it may be remembered that the rhodium porphyrin carbene complex $[RhCR_2(TPP)(CNCH_2Ph)]PF_6$ ($R = NHCH_2Ph$) (Table 2) crystallizes with an organometallic ligand *trans* to the carbene ligand [175].

Nucleophilic additions to alkylrhodium(III) porphyrins – An interesting reaction sequence (31) was observed when RhMe(OEP) was treated with an excess of lithium aryls LiAr [324].

$$\mathrm{RhMe(OEP)} \xrightarrow{(a)} \underset{\text{brown}}{[\mathrm{RhAr_2(OEP)}]^-} \xrightarrow{(b)} \underset{\text{green}}{[\mathrm{RhAr(OEPAr)}]^-} \xrightarrow{(c)} \underset{\text{orange}}{\mathrm{RhAr(OEPAr)}} \quad (32)$$

Conditions: (a) LiAr/THF (15-fold excess); (b) 5 h standing; (c) NH_4Cl, H_2O, O_2.

Step (a) of sequence (31) implied replacement of the rhodium bound methylide by an aryl anion and addition of a second one *trans* to the first one. This brown species displayed a hyper spectrum like the bis(dithiolato)rhodium(III) porphyrin anions $[Rh(SR)_2(OEP)]^-$ mentioned in Sect. 3.3 (Table 6, path j). Step (b) involved a slow rearrangement of one of the aryl anions which was added to a *meso* (C-5) position of the OEP ligand causing a phlorin-like spectrum with λ_{max} at about 438 and 805 nm, as found with a previously described aluminium phlorinate anion $[AlR(OEPH)]^-$ (R = *i*Bu) formed from $[AlH(iBu)_2]_2$ and $H_2(OEP)$ [325]. Therefore, it was concluded than an anionic aryl rhodium(III) complex of the 5-aryloctaethylphlorinate trianion was formed. In step (c), this species was quenched with aqueous ammonium chloride and air; thus, the phlorin anion was reoxidized to the orange *meso*-aryl substituted octaethylporphyrinato arylrhodium(III) complex RhAr(OEPAr). This rare nucleophilic addition is obviously promoted by the rather high (+3) positive charge of the Al(III) or Rh(III) central ion.

Reactions of rhodium(III) porphyrins with olefins and acetylenes – Ogoshi et al. [326] have described the reactions of vinyl ether with rhodium (III) porphyrins which are depicted in reaction sequence (33). Step (a) appears to be an insertion of the olefin into the Rh–Cl bond followed by alcoholysis of a chlorosemiacetal to the acetal, step (b) is the hydrolysis of the acetal to the aldehyde. The insertion is thought to start by heterolysis of the Rh–Cl bond producing a cationic species which forms a π-complex with the electron-rich olefin.

$$\mathrm{RhCl(OEP)(H_2O)} \xrightarrow{(a)} \mathrm{Rh(CH_2CH\{OEt\}_2)(OEP)} \xrightarrow{(b)} \mathrm{Rh(CH_2CHO)(OEP)} \quad (33)$$

Conditions: (a) $CH_2 = CHOEt$, EtOH, NEt_3; (b) silica gel or water in chloroform.

The acetal was likewise prepared from chloroacetal and $[Rh(OEP)]^-$, the aldehyde from $[Rh(OEP)]_2$ and ethyl vinyl ether [326].

The unsaturated system of arenes may also be directly attacked by Rh(III) porphyrins when their electrophilicity is enhanced by offering just weakly coordinating anions. Thus, treatment of RhCl(OEP) with $AgClO_4$ or $AgBF_4$ in neat benzene derivatives C_6H_5R (R = H, Me, OMe, Cl) yielded the corresponding arylrhodium(III) porphyrins Rh(p-C_6H_4R) (OEP) [59, 327]. These compounds had been independently synthesized from the corresponding lithium aryls and RhCl(OEP) [54]. Photolysis of these arylrhodium(III) species yielded the corresponding biphenyls [59].

The reaction of acetone with RhCl(OEP) in the presence of $AgClO_4$ or with Rh(OEP) ClO_4 yielded the acetylmethylrhodium(III) complex Rh(CH_2COMe) (OEP) [283, 328]; this reaction could be viewed as having started from the enol form of acetone.

Organometallic reactions of hydridorhodium or -iridium porphyrins – The various reactions leading to organometallic derivatives of rhodium or iridium porphyrins have already been depicted in Scheme 3 and Table 6 (paths l, m, n, – r, s). In the following paragraphs, some especially noteworthy reactions will be described. At first, the ones starting from hydrides will be mentioned. Due to the existence of the reaction paths q and p (Scheme 3), it is not always clear whether a reaction really starts from the hydridometal or the dimeric metal porphyrin species. Most of the contributions to this field came from Wayland et al. starting with the observation that hydricorhodium(III) complexes gave labile formylrhodium(III) species, e.g. Rh(CHO)(OEP), according to sequence (34) [329]. Most of the investigations have been done with OEP complexes, but the corresponding TPP complexes have been shown to react very similarly to the OEP complexes [330].

$$[\mathrm{Rh(OEP)}]_2 \xrightarrow{\mathrm{H_2}} 2\ \mathrm{RhH(OEP)} \xrightarrow{\mathrm{CO}} 2\ \mathrm{Rh(CHO)(OEP)} \qquad (34)$$

The hydride was not only formed from the dimer $[Rh(OEP)]_2$, but likewise by a base reaction of RhCl(OEP)CO with solid KOH; thus the formyl complex Rh(CHO) (OEP) was simply prepared from RhCl(OEP)CO, excess CO and KOH [331]. A crystal structure determination of Rh(CHO)(OEP) proved its existence in coordination type **B** [332]. It is remarkable that IrH(OEP) does not form a formyl complex with CO, but rather adds the CO ligand *trans* to the hydrido ligand, giving IrH(OEP)CO [333].

RhH(OEP) reacted reversibly with acetaldehyde, MeCHO, and other aldehydes under formation of α-hydroxyalkylrhodium(III) porphyrins, e.g. Rh(CHMeOH)(OEP). By condensation, a symmetrical ether was formed from this compound which in turn was subject to a hydride transfer disproportionation, yielding a mixture of acetylrhodium(III) and ethylrhodium(III) species, Rh(COMe)(OEP) and RhEt(OEP) [334]. Action of formaldehyde on RhH(OEP) yielded the hydroxymethyl derivative Rh(CH_2OH)(OEP) which was also prepared by reduction of the formyl compound, Rh(CHO)(OEP), with $[AlH(iBu)_2]_2$. The hydroxymethyl compound reacted with RhH(OEP) according to sequence (35), yielding methanol, thus presenting a reaction thought to be important for the hydrogenation of carbon monoxide [335]. This sequence

$$Rh(CH_2OH)(OEP) + RhH(OEP) \rightarrow [Rh(OEP)]_2 + MeOH \qquad (35)$$

came at the end of a series of reactions which had started with the equilibration of $[Rh(OEP)]_2$ with H_2 and CO, yielding the formyl rhodium(III) derivative Rh(CHO)(OEP). Photolysis of the latter in presence of RhH(P) yielded small stationary concentrations of formaldehyde which vanished after turning off the light because of the reaction (36) [336] which is nothing else than the already

$$[Rh(OEP)]_2 + CH_2O \rightarrow RhH(OEP) + Rh(CHO)(OEP) \qquad (36)$$

mentioned attack of the dimeric rhodium(II) porphyrins on CH bonds. The reaction of RhH(OEP) with glycolaldehyde likewise resulted in an addition to the aldehyde group. The 1,2-dihydroxyethylrhodium(III) porphyrin thus formed lost a molecule of water in boiling benzene, yielding a formylmethylrhodium(III) porphyrin, $Rh(CH_2CHO)(OEP)$ [337].

Organometallic reactions of dimeric rhodium(II) or iridium(II) porphyrins – The review of Guilard, Kadish, and coworkers [306] has already been cited. This gives a clear evaluation of the reactions of $[Rh(OEP)]_2$ and $[Ir(OEP)]_2$ with a variety of organic substrates as far as described up to 1987. The important reactions with aliphatic and benzylic CH bonds have already been mentioned in Sect. 3.3, see Eq. (20). Here, some more recent developments, especially concerning the reactions with CO or olefins, will be elaborated.

Halpern et al. had already stressed the importance of radical mechanisms in the oxidative addition and insertion reactions of both $[Rh(OEP)]_2$ and Rh(OEP) [338]. Thus, $[Rh(OEP)]_2$ reacted with trimethylphosphite according to sequence (37), forming a rhodiophosphonate $Rh(PO\{OMe\}_2)$ (OEP) and methyl radicals which were subject to further reactions [339]. In the presence of excess $P(OMe)_3$, they were trapped by formation of $MePO(OMe)_2$ in more than stoichiometric quantities, indicating a radical chain process.

$$[Rh(OEP)]_2 + 2\ P(OMe)_3 \rightarrow 2\ Rh(PO\{OMe\}_2)\ (OEP) + 2\ CH_3^{\bullet} \qquad (37)$$

Reactions of monomeric and dimeric rhodium(II) porphyrins with carbon monoxide – As already reported in Sect. 3.3, a carbonylrhodium(II) porphyrin behaves as an acyl radical. Hence, if possible, dimerization or coupling reactions occur. Evidence for the formation of isomeric 2 : 1 Rh(P) : CO adducts, namely a mono-adduct of the dimer and a metallo ketone complex, and a dimeric 1 : 1 adduct in the reaction of $[Rh(OEP)]_2$ with carbon monoxide according to sequences (38) and (39) has been presented [340, 341]; solution equilibria and structures have been studied essentially by ^1H NMR, ^{13}C NMR, and IR spectroscopy. The first half of sequence (38) and reaction (39) occurred in parallel at CO pressures up to 12 atm at 297 K. At higher pressures, or at lower temperatures, the double-insertion of CO shown in the last step of (38) was observed.

$$[Rh(OEP)]_2 \xrightarrow{CO} (OEP)RhC(O)Rh(OEP)$$

$$\xrightarrow{CO} (OEP)RhC(O)C(O)Rh(OEP) \quad (38)$$

$$[Rh(OEP)]_2 \xrightarrow{CO} (OEP)RhRh(CO)(OEP) \quad (39)$$

For Rh(TMP), the reaction (40) with carbon monoxide yields a diamagnetic C–C bound dicarbonyl dimer which is in equilibrium with a small amount of the paramagnetic monocarbonyl [271a].

$$2\ Rh(TMP) + 2\ CO \rightarrow 2\ RhCO(TMP)$$

$$\leftrightarrows (TMP)RhC(O)C(O)Rh(TMP) \quad (40)$$

Only a monocarbonyl was formed when a most sterically demanding porphyrin complex, Rh(TTiPP), was subjected to carbonylation.

Formation of olefin complexes – Dimeric metalloporphyrins $[M(P)]_2$ react with olefins $CH_2{=}CHR$ (M = Rh, R = H [342], Ph [338], Me [343], COOEt [344]; M = Ir, R = OEt [272]) to form two-carbon alkyl bridged complexes according to Eq. (41) (see Sect. 3.3). Step (a) occurs for the dimers formed from sterically unencumbered porphyrins.

$$[M(P)]_2 \underset{(a)}{\leftrightarrows} 2\ M(P) \underset{(b)}{\xrightarrow{CH_2=CHR}} (P)MCH_2CH(R)M(P) \quad (41)$$

For sterically hindered porphyrins like Rh(TMP), the reaction starts with step (b). If the sterical hindrance is increased as in Rh(TTEPP), the reaction with ethene proceeds along sequence (42).

$$2\,M(P) + 2\ CH_2{=}CH_2 \underset{(a)}{\leftrightarrows} 2\,M(CH_2{=}CH_2)\,(P) \underset{(b)}{\leftrightarrows} (P)\,M(\{CH_2\}_4)M(P) \quad (42)$$

The quenching of the radical formed along step (a) in sequence (42) cannot occur by uptake of a second M(P) radical, but rather by its dimerization according to step (b). If the extremely hindered porphyrin ligand in monomeric Rh(TTiPP) was involved in sequence (42), the coupling shown in step (b) was considerably slowed down.

Thus, a 1 : 1 Rh(TTiPP)-ethene π-complex could be frozen out. The ^{1}H and ^{13}C NMR spectra served to assign the structures of the diamagnetic species. The paramagnetic $Rh(CH_2{=}CH_2)(TTiPPP)$ was identified in a toluene glass and its structure elucidated by its ESR spectrum using $^{13}CH_2{=}^{13}CH_2$ as the π-ligand. The symmetrical nature of the complex was proven by observation of a doublet of triplets arising from nuclear hyperfine coupling with ^{103}Rh and two equivalent ^{13}C nuclei in ethene.

Substituted propenes $CH_2{=}CHCH_2R$, namely allylbenzene, allyl cyanide, and hexene-1 (R = Ph, CN, n-Pr), were shown to yield stable σ-alkylrhodium (III) porphyrins in moderate yields according to reaction (43) [269]. An isomerization and an attack on a saturated CH bond as shown in Eq. (20) could be the reason for the outcome of this type of reaction which is different to the examples shown in Eqs. (41) and (42).

$$[Rh(OEP)]_2 + 2\ CH_2{=}CHCH_2R \rightarrow 2\ Rh(CH_2CH{=}CHR)(P) \qquad (43)$$

Metalloradical reactions with acrylates have also been observed [344]. The sterically unhindered dimer $[Rh(OEP]_2$ "incorporated" and thus reduced acrylic acid derivatives, $CH_2{=}CHCOOR$ (R = H, Me, Et), to form bridged systems $(OEP)Rh(CH\{COOR\}CH_2)Rh(OEP)$, similarly to step (41b). The sterically hindered monomeric Rh(TMP) inhibits the formation of these two-carbon alkyl bridged complexes, but rather promotes alkene coupling of acrylates, yielding four-carbon alkyl bridged complexes, namely $(TMP)Rh(CH_2CH\{COOR\}$-$CH\{COOR\}CH_2)Rh(TMP)$, as in step (39b). Methyl methacrylate, however, underwent an attack on the methyl group as described for toluene or methane given in sequence (20) as already mentioned in Sect. 3.3 (Scheme 3, paths − q, − p, − r). The radical-like rhodium(II) porphyrins were ineffective at initiation of thermal polymerization of acrylates, however, Rh(TMP) catalyzed a photo-promoted polymerization that had living character. The radical-like character of $[Rh(OEP)]_2$ was also evident in its reactions with silanes or stannanes which are thought to proceed according to reaction (44), yielding trialkylsilyl- or trialkylstannylrhodium(III) porphyrins. $RhSiMe_3(OEP)$ was crystallographically confirmed [345].

$$[Rh(OEP)]_2 + 2\ SiHEt_3 \rightarrow 2\ RhSiMe_3(OEP) + H_2 \qquad (44)$$

Some electrochemically initiated organometallic reactions of rhodium and iridium porphyrins have been explored and reviewed by Kadish et al. [178, 306]. For more recent papers concerning this matter, the reader is referred to Sect. 5.

5 Further Chemical Reactions of Noble Metal Porphyrins: Notes on Electrochemistry, Catalysis, and Other Applications

The enormous research activity devoted to metalloporphyrins, and in special, to the noble metal porphyrins is governed by the idea of biomimesis, i.e. the desire of chemists to understand and imitate nature with its ingenious biocatalytic pathways. In this respect, two important neighbouring disciplines to synthetic chemistry will be touched on in this article: electrochemistry (Sect. 5.1) and catalysis (Sect. 5.2). However, in order to keep the article within certain limits, only some notes will be given on these fields, which sometimes are linked by the term "electrocatalysis" and bear some relevance to material science. Further chemical reactions not described in this article may be found in the articles cited in the following tables. Some special applications will be added (Sect. 5.3).

5.1 Electrochemical Investigations of Noble Metal Porphyrins

A thorough electrochemical characterization of new metalloporphyrins is nowadays state of the art for the synthetic inorganic chemist. In many of the papers cited in Sects. 3 and 4, a characterization of the new complexes by cyclic voltametry and electrolysis at controlled potential has been done. Thin-layer spectroelectrochemistry is very fruitful [346]. Fortunately, apart from classical articles of Davis et al. [347], Felton et al. [292], Fuhrhop et al. [293], Buchler et al. [190], more recent reviews of Kadish et al. are available which systematically cover the field of general metalloporphyrins [294] or organometallic porphyrin complexes [306]. Therefore, a short, update of these articles will be given in the form of Table 7. For details, the reader is referred to the original literature.

Table 7. Publications on electrochemical investigations of noble metal porphyrins

Complex type	Key words	References
$M(OEP)_2$	Metalloporphyrin dimers (M = Ru, Os)	[348]
RuLL′ (ToAPP)	Electrocatalysis of O_2 reduction	[349]
$Ru_2(TBP)LL'$	Cofacial Bis(metallo)diporphyrins	[225]
RuCO(TPP)L	Carbonylruthenium(II) porphyrins	[350]
RuCO(TPP)L	Ring methylation of $[RuCO(TPP)1]^-$	[351]
RuR(P)	Organoruthenium(III) porphyrins	[308]
$RuR_2(P)$	Diorganoruthenium(IV) porphyrins	[309, 310]
$OsX_2(P)$	Osmium(IV) porphyrins (redox chemistry)	[250]
$Rh(O_2)(TPP)$	Electroreduction of a dioxygen adduct	[352]
$[RhL(TPP)L']PF_6$	Phosphine and carbene complexes	[353]
$[Rh(CO)_2]_2(P)$	Oxidative rhodium insertion	[354]
RhCl(Pc)Py	Reductive deligation and dimerization	[355]
$[Ir(CO)_2]_2(TPP)$	Oxidative iridium insertion	[356]
IrCl(OEP)CO	Porphyrin-centered one-electron oxidation	[346]
M(TPrPyC)	Metallotetrapropylporphycenes (M = Pd, Pt)	[357]
$M(C_4\{OEPyl\}_2)$	Study of diacetylene-bridged diporphyrin[a]	[358]
$PtCl_2(P)$	Irreversible two electron reduction to Pt(P)	[85]
Ag(TPP)	Ag(I), Ag(II), Ag(III) tetraarylporphyrins	[359]
$Ag(TCl_8PP)$	Exists as Ag^{3+}/π-radical trianion ($P^{\cdot 3-}$	[360]

[a] $H_2(C_4\{OEPyl\}_2)$: 1,4-Di(octaethylporphyrin-5-yl)buta-1,3-diyne; M = Pd, Pt

5.2 Catalytic Activity of Noble Metal Porphyrins

Fascinated by the oxygen turnover processes catalyzed by heme proteins, namely the peroxidases, cytochromes P-450, and catalase, thousands of researchers appear be looking for active and robust metalloporphyrin catalysts.

This field has been reviewed several times [15, 16, 361]. The active species are thought to be oxometal porphyrins in which the metals are in oxidation states higher than the usual ones, and the most active oxo complexes are found in the 3d transition metal series, e.g. FeO(P) or $[FeO(P)]^+$, MnO(P) or $[MnO(P)]^+$, CrO(P) or $[CrO(P)]^+$. Since the existence of oxometal porphyrins in which the metal assumes a "high", i.e. greater than + 4, oxidation state is restricted to the metals ruthenium and osmium (see Sect. 3), it is not surprising that oxygenation catalysts have only been looked for in the set of oxoruthenium or oxoosmium porphyrins. Usually, these catalytic reactions involve a source of active oxygen (dioxygen, hydrogen peroxide, iodosylbenzene, hypochlorite, peroxosulfate, amine-N-oxides, etc.), the organic substrate, and the catalyst. The catalyst (in its "resting state") is administered with its metal in a "normal" (+ 2 or + 3) oxidation state and transformed to a reactive oxometal complex in an elevated oxidation state (+ 4, + 5, + 6) which transfers an oxygen atom to the substrate, effecting in most cases either an addition or an insertion to one or two sp^2-C atoms or into a C–H bond.

A study with $OsBr(TPP)PPh_3$ as a resting state and iodosylbenzene as the oxidant was not very promising [239]. Furthermore, dioxoosmium(VI) porphyrins $OsO_2(P)$, despite their record in oxidation state of a metal within the porphyrin, are only weak oxidants.

Since RuO_4 has a much higher oxidation potential than OsO_4, it could be expected that $RuO_2(P)$ might be more useful for oxidations. As has been described by Groves et al., this is indeed the case. After the synthesis of $RuO_2(TMP)$ [243], it was discovered that this compound was a competent stoichiometric oxidant as shown in sequence (45) and an active catalyst for the epoxidation of olefins with dioxygen [362]. The catalytic cycle is composed of the reactions (46), (47), and (48). It was applied to cyclooctene, norbornene (C_7H_{10}), and *cis*- and *trans*-β-methylstyrene; the latter compounds were used to demonstrate that the epoxidation proceeded with retention of configuration.

$$RuO_2(TMP) + 2\ C_7H_{10} + 2\ Py \rightarrow RuPy_2TMP + 2\ C_7H_{10}O \qquad (45)$$

$$RuO_2(TMP) + \text{olefin} \rightarrow RuO(TMP) + \text{epoxide} \qquad (46)$$

$$RuO(TMP) + RuO(TMP) \rightarrow RuO_2(TMP) + Ru(TMP) \qquad (47)$$

$$Ru(TMP) + 1/2O_2 \rightarrow RuO(TMP) \qquad (48)$$

The catalytic cycle lives on the fact that Ru(II) porphyrins are autoxidized to dioxoruthenium(VI) porphyrins in the presence of just weak donor ligands like THF or benzene.

The *ortho* buttressing of the TMP ligand was found to be essential for catalytic activity. Thus, the tetra(*p*-tolyl) porphyrin $Ru(TTP)(THF)_2$, which is known to form an inactive μ-oxodiruthenium(IV) complex upon oxygenation, was inactive as an oxygenation catalyst, as were RuCO(TMP)L and $Ru[P(OMe)_3]_2(TMP)$. This means that the reaction medium must not contain π-acceptor ligands which prevent an autoxidation of Ru(II) porphyrins.

A dihydroxoruthenium(IV) porphyrin, $Ru(OH)_2(TMP)$, was claimed to have been prepared according to reaction (49) from EtOH and to be a more active catalyst than $RuO_2(TMP)$ [363]. The structure of the related bis(isopropoxy)-ruthenium (IV) complex $Ru(OiPr)_2(TMP)$ was confirmed by X-ray crystallography. Its formation from $RuO_2(TMP)$ was explained by reactions (49) and (50) [364].

$$RuO_2(TMP) + iPrOH \rightarrow Ru(OH)_2(TMP) + Me_2CO \tag{49}$$

$$Ru(OH)_2(TMP) + 2\ iPrOH \rightarrow Ru(OiPr)_2(TMP) + 2\ H_2O \tag{50}$$

Coordinatively labile ruthenium(II) porphyrins $Ru(P)(THF)_2$ (P = TTP, TMP) catalyse the *cis- trans* isomerization of epoxides under mild conditions, probably by coordination of the epoxide and ring opening via a carbon radical [365]. The lifetime of the catalysts is restricted due to carbon monoxide abstraction from coordinated epoxide to yield inactive carbonylruthenium(II) complexes, e.g. RuCO(TMP)THF [366].

These important observations have stimulated further research in this field. Heteroaromatic N-oxides have been used as oxygen sources in oxidations of alkanes (e.g. adamantane) and alkyl alcohols catalyzed by $RuO_2(TMP)$ [367]. The latter compound stoichiometrically transformed alkyl thioethers R_2S to the corresponding sulfoxides SOR_2. This reaction became catalytic in the presence of dioxygen, but the active Ru(IV) or Ru(II) intermediates which regenerate the Ru(VI) species according to reactions (47) and (48) are slowly blocked by formation of inert $Ru(SOR_2)_2(TMP)$ in which the sulfoxide ligands are bound via the sulfur atom to the Ru(II) center [368].

A very interesting application of the "Groves-Quinn oxidation" depicted in Eqs. (46) to (48) to unsaturated steroids was described in a series of papers by Marchon, Ramasseul et al. [369]. Due to their asymmetric shape steroids give a good chance of deriving the stereochemical requirements of their reactions. The variable reactivity and specificity of the epoxidation process performed with a variety of steroidal olefins was explained by assuming a transition state in which the steroid core approaches the oxoruthenium bond approximately from a direction perpendicular to the porphyrin ring. Such a transition state geometry provides a good fit between the porphyrin catalyst and the steroid substrate when the β-side of the latter faces the oxo ligand and explains the β-stereospecificity of the catalytic epoxidation of cholesterol acetate.

The steric requirements of the tetramesityl porphyrin ligand are likewise determining the optimal geometry of the transition states in these reactions. Therefore, it was consequent to look for a chiral porphyrin ligand which could add enantioselectivity to the processes already described. Simonneaux et al. [370] have synthesized a set of chiral porphyrins derived from the four atropisomers of tetrakis(*o*-aminophenyl) porphyrin, $H_2(ToAPP)$, by acylation with (R)-(+)-α-methoxy-α-trifluormethylphenylacetyl chloride. This mixture of chiral porphyrins $H_2(P^*)$ was metallated with $Ru_3(CO)_{12}$ in *o*-dichlorobenzene and the resulting chiral isomers of RuCO(P*)THF separated by thin layer chromatography. The α, β, α, β isomer of RuCO(P*)THF was complexed in a highly stereoselective manner (> 95%) with racemic benzylmethylphenylphosphine P^*R_3 to form one of the possible product diastereoisomers, $RuCO(P^*)P^*R_3$ [370b]. The α, β, α, β isomer of the dioxoruthenium(VI) porphyrin, $RuO_2(P^*)$, was also made and found to perform asymmetric induction in the oxygenation of racemic P^*R_3 to form the oxophosphorane (S)-(+)-OP^*R_3 in 41% enantiomeric excess and retention of the configuration at the phosphorus atom [370c].

Attempts to prepare chiral rhodium porphyrins are also a challenge in view of the many catalytic organometallic reactions alluded to in the papers cited in Sect. 4.2. H_2(ToAPP) was reacted with camphanic acid chloride to obtain an α, β, α, β-tetrakis(*o*-camphanylamidophenyl)porphyrin, $H_2(P^*)$, which after transformation to its chloromanganese(III) complex MnCl(P*) only caused small enantiomeric excesses of chiral epoxides when catalyzing the epoxidation of styrene. The stereochemistry of this porphyrin ligand was elucidated after its complexation with $[RhCl(CO)_2]_2$ in form of RhCl(P*) using ^{13}C and ^{1}H resonance spectra. It was concluded that chiral appendages with a lower degree of flexibility than that of the amidic bridge and without coordinating substituents might give better results [371].

Some other catalytic events prompted by rhodium or ruthenium porphyrins are the following: 1. Activation and catalytic aldol condensation of ketones with $Rh(OEP)ClO_4$ under neutral and mild conditions [372]. 2. Anti-Markovnikov hydration of olefins with $NaBH_4$ and O_2 in THF, a catalytic modification of hydroboration-oxidation of olefins, as exemplified by the one-pot conversion of 1-methylcyclohexene to (*E*)-2-methylcyclohexanol with 100% regioselectivity and up to 90% stereoselectivity [373]. 3. Photocatalytic liquid-phase dehydrogenation of cyclohexanol in the presence of RhCl(TPP) [374]. 4. Catalysis of the water gas shift reaction in water at 100 °C and 1 atm CO by $[RuCO(TPPS_4)H_2O]^{4-}$ [375]. 5. Oxygen reduction catalyzed by carbon supported iridium chelates [376]. – Certainly these notes can only be hints of what can be expected from new noble metal porphyrin catalysts in the near future.

5.3 Further Applications of Noble Metal Porphyrins

At the end of our trip through the realm of noble metal porphyrins, a glance at some biochemically or biophysically relevant new derivatives may be appropriate. Most of the complexes belong to a class of porphyrins which was not a subject of systematic examination by the authors: *peripheral* metalloporphyrins, in which the metal is not bound to the four central nitrogen atoms of the pyrrole rings, but rather to the side chains of macrocycles.

Biochemically important platinum complexes. – Since the mechanism of action of anticancer drugs like *cis*-platin, *cis*-$PtCl_2(NH_3)_2$, implies *N*-guanosine-binding to two nucleobases in a polydesoxyribonucleotide chain, it cannot be expected that a "central" platinum porphyrin shows any cancerostatic activity. Nevertheless, a variety of "peripheral" platinum complexes have been prepared from natural porphyrins, the two (13, 17) propionic acid side chains of which were incorporated into *cis*-dicarboxylatodiammineplatinum(II) moieties or transformed into two 3-aminopropyl groups that were used to bind a *cis*-$PtCl_2$ fragment. These Pt complexes were hoped to show the combined capability of the porphyrins to be accumulated by cancer cells and the *cis*-diammine platinum fragment to stop tumor growth, and indeed, a notable antitumor activity of the diamminedicarboxylatoplatinum(II) complexes was found in vitro (MDA-MB 231 mammary carcinoma cell line) and in vivo (three malign tumors of the mouse) [377]. On the other hand, the interaction of a series of water-soluble cationic "central" palladium(II) porphyrins, i.e. $[Pd(TMPyP)]^{4+}$ and its *o*- and *m*-pyridyl isomers, with desoxyribonucleic acid (DNA) or synthetic polynucleotides was studied [378]. Both intercalation between base pairs (*m*- and *p*-pyridyl isomers) and electrostatic binding to the phosphate chain (*o*-isomer) was observed. Photophysical investigations of the adducts in the presence of intercalated acridine orange led to the conclusion that the DNA duplex does not provide a particularly attractive medium for electron exchange.

Biophysical Aspects: *Chromophore Aggregates* – The charge separation in photosynthetic model compounds requires aggregates of chromophores that can act as donors, sensitizers, and acceptors in light-driven charge separation. The axial coordination sites of Ru(II) or Os(II) porphyrins have been used to link these chromophores to other redox active centers via coordinative bonds. A naphthoquinone unit was fixed via two 2,3-linked spirane rings to a thioether function which served as an axial ligand to an OsCO(TTP) fragment. Thus, a naphthoquinone–osmiumporphyrin aggregate was formed. Emission and transient absorption spectra, however, did not shown any charge separation occuring upon illumination [379]. The complexes $Mo(NO)\{HB(dmpz)_3\}$-(Cl)L $[HB(dmpz)^-$: tris(3,5-dimethylpyrazolyl) borate] and RuCO(TPP)L or $Ru(THF)_2(TPP)$ were connected at the positions of L and L, respectively, by bridging ligands L ∧ L, 4,4′-bipyridyl, to give the binuclear or trinuclear complexes Mo(NO) $\{HB\text{-}(dmpz)_3\}$(Cl) L ∧ LRuCO(TPP) or [Mo(NO)-$\{HB(dmpz)_3\}$(Cl) L ∧ $L]_2Ru(TPP)$. The complexes were identified inter alia by fast atom bombardment mass spectra and cyclic voltammetry; the measured redox potentials showed only weak interactions between the connected coordination centers [380].

Not only the "axial" coordination of "central" metalloporphyrins may be used to form aggregates, but likewise, the metallation of unidentate coordination sites of porphyrin substituents may lead to oligomeric "peripheral" metalloporphyrins (see also Sect. 2.1.1). 5,15-Bis(ethinyl)-2,8,12,18-tetraethyl-3,7,13,17-tetramethylporphyrinatozinc(II) was coupled with *trans*-$Pt(PEt_3)_2Cl_2$ in diethylamine using Cu(I) catalysis to give, amongst other oligomers and fragments, a cyclic trisporphyrin in 16% yield [381]. In this trisporphyrin, the meta-phenyl positions are linked via ethynyl-*cis*-bis(triethylphosphine)platinum(II)-ethynyl bridges to form a cavity in which the three porphyrin planes are incorporated to the rectangular faces of a trigonal prism. The three zinc ions of the trisporphyrin and their acceptor sites are located in a large triangle to which a suitable guest molecule was fitted: tris[3-(4-pyridyl)pentane-2,4-dionato]aluminium(III). This complex served as a tridentate ligand and was bound to the three zinc(II) ions with an association constant the lower limit of which was estimated by UV spectroscopy as 1×10^{10} $dm^3 mol^{-1}$.

The complexes *trans*-bis(pyridine)dichloropalladium(II) and *cis*-bis-(pyridine)dichloroplatinum(II) served as an orientation in the second study [382]. Vicinal 5,10-bis(4-pyridyl)-15,20-diphenylporphyrinatozinc(II) or transversal 5,15-bis(4-pyridyl)-10,20-diphenylporphyrinatozinc(II) were treated in toluene with *trans*-$PdCl_2(NCPh)_2$ or *cis*-$PtCl_2(NCPh)_2$, respectively, yielding, among zig-zag configurated oligomers, square-shaped tetranuclear Pd or Pt complexes in which the vicinal or transversal bis(pyridyl) porphyrins were linked by *trans*-$PdCl_2$- or *cis*-$PtCl_2$-fragments, respectively. In the resulting square tetraporphyrins either the zinc ions were sitting in the corners of a square (*trans*-$PdCl_2$ units on the edges), or the Pt ions of the *cis*-$PtCl_2$ moiety (zinc ions on the edges). The tetranuclear zinc-free species were secured by NMR and electrospray mass spectroscopies. The photophysical study of these porphyrin tetrads is under way.

This supramolecular outlook to large cyclic oligoporphyrins held together by "peripheral" metal coordination may conclude this stroll through the world of noble metal porphyrins. The authors have left the shape of these large molecules to the readers imagination because of the large drawings that would have been necessary to show the formulas.

6 Epilogue – The Special Character of Noble Metal Porphyrin Entities

As compared with the corresponding 3d metal ions, the 4d or 5d noble metal ions have something special when coordinated to the porphyrin ligand. Although covalent metal-to-ligand bonds become much stronger when descending within a group, axial ligand exchange in general remains rather rapid, especially in axially unsymmtrical entities, such as L–M–CO or X–M$\equiv$N systems in which the ligands trans to the axial carbonyl or nitride are substitutionally labile, due to the trans effect of these latter strongly π-bonding ligands [190]. The porphyrin ligand appears to squeeze equatorially the soft noble metal acceptor ions which are found right in the porphyrin ring plane in the axially symmetrically ligated complexes although the size of the ions as given by ionic radii would not lead to such an expectation. Thus, d_π backbonding increases considerably in the series of configurations $3d^6 < 4d^6 < 5d^6$. Olefin complexation thus is observed at Ru(II) and Os(II) centers, but oxidative addition does not take place because the robust equatorial porphyrin ligand does not allow the liberation of *cis*-coordination sites. Only the dimeric rhodium(II) porphyrins behave like complexes with two adjacent coordination sites and initiate a variety of organometallic reactions via radical intermediates; here most promising developments in catalysis are to be expected. The equatorial metal-to-porphyrin π-backbonding free or any axial ligand donor capacity can be evaluated with Pd and Pt porphyrins. Thus, the special role of noble metal porphyrins may be summarized as follows: high electron density in axial direction with lack of cis coordination sites and robust equatorial coordination frame provide a special reactivity not found with other metal catalysts.

Acknowledgements. Without long-term financial support from the Deutsche Forschungsgemeinschaft, the Fonds der Chemischen Industrie, Degussa AG (gifts of noble metal carriers for metal insertions) and Bayer AG (gifts of solvents) the results which have been elaborated in the laboratories at Aachen and Darmstadt and cited here could not have been achieved. The authors thank all their colleagues and team-mates at the Institut für Anorganische Chemie during the cited work. Final assistance of Dipl.-Ing. Matthias Mehler during the completion of the manuscript is gratefully acknowledged.

7 References

1. Remy H (1957) Lehrbuch der Anorganischen Chemie, 9. Aufl., Bd. I, Akademische Verlagsgesellschaft, Leipzig, p 818
2. Ühlein E (1978) Römpps Chemisches Wörterbuch, Franck'sche Verlagsbuchhandlung, Stuttgart, p 196
3. Griffith WP (1967) The chemistry of the rarer platinum metals (Os, Ru, Ir, and Rh), Wiley-Interscience, London
4. Wiberg N (1985) Holleman-Wiberg, Lehrbuch der Anorganischen Chemie, de Gruyter, Berlin, p 1186
5. Greenwood NN, Earnshaw A (1984) Chemistry of the elements, Pergamon, Oxford, p 30
6. Cotton FA, Wilkinson G (1980) Advanced inorganic chemistry, 4th edn. Wiley-Interscience, New York, p 966
7. Buchler JW (1975) In: Smith KM (ed) Porphyrins and metalloporphyrins. Elsevier, Amsterdam, p 157
8. Buchler JW (1978) In: Dolphin D (ed) The porphyrins, Academic Press, New York, Vol. I, p 390
9. Bohrer R, Kalbskopf B, Richter H-J, Ditten L, Leichner L, Best E, "GABCOM & GABMET; Abkürzungen von Verbindungen und Methoden aus Chemie und Physik", Hrsg.: Gmelin-Institut für Anorganische Chemie, Springer-Verlag, Berlin 1993
10. Liebscher W, Neels J (1994) "Nomenklatur der Anorganischen Chemie", Hrsg.: Gesellschaft Deutscher Chemiker, VCH-Verlag, Weinheim
11. Hudson MF, Smith KM (1975) Tetrahedron 31: 3077
12. Buchler JW, Kruppa S (1990) Z. Naturforsch 45b: 518
13. Collman JP, Garner JM, Kim K, Ibers JA (1988) Inorg Chem 27: 4513
14. Ibers JA, Holm RH (1980) Science 209: 223
15. McMurry TJ, Groves JT (1986) In: Ortiz de Montellano (ed) Cytochrome P 450: Structure, Mechanism, and Biochemistry, Plenum, New York, pp 1–28
16. (a) Meunier B (1983) Bull Chim Soc France 1983, II: 345;
    (b) Meunier B (1986) Bull Chim Soc France p. 578
    (c) Meunier B (1992) Chem Rev 92: 1411

17. Mochida I, Fujitsu H (1984) Shokubai 26: 443
18. Yoshida Z (1984) Heterocycles 21: 331
19. Morgan B, Dolphin D (1987) Struct Bonding [Berlin] 64: 115.
20. Kitagawa T, Ozaki Y (1987) Struct. Bonding [Berlin] 64: 71
21. Anderson LA, Dawson JH (1991) Struct Bonding [Berlin] 74: 1
22. Buchler JW (1987) Comments Inorg Chem 6: 175
23. Sima J (1995) Struct Bonding [Berlin] this volume
24. Ibers JA, Pace LJ, Martinsen J, Hoffman BM (1982) Struct Bonding [Berlin] 50: 1
25. (a) Hanack M (1982) Naturwiss 69: 266
    (b) Hanack M (1985) Israel J Chem 25: 205
26. Tsutsui M, Kasuga K (1980) Coord Chem Rev 32: 67
27. Wöhrle D, Meyer G (1985) Kontakte (Darmstadt) p 38
28. Wöhrle D (1986) Kontakte (Darmstadt) p 24
29. Ostfeld D, Tsutsui M (1974) Acc Chem Res 7: 52
30. Fleischer EB, Thorp R, Venerable D (1969) J Chem Soc Chem. Commun. p 475
31. Chow BC, Cohen IA (1971) Bioinorg Chem 1: 57
32. Hopf FR, O'Brien TP, Scheidt WR, Whitten DG (1975) J Am Chem Soc 97: 277
33. Tsutsui M, Ostfeld D, Francis JN, Hoffman LM (1971) J Cord Chem 1: 115
34. Barley M, Becker JY, Domazetis G, Dolphin D, James BR (1983) Can J Chem 61: 2389
35. Rillema DP, Nagle JK, Barringer LF, Meyer TJ (1981) J Am Chem Soc 103: 56
36. Ariel S, Dolphin D, Domazetis G, James BR, Leung TW, Rettig SJ, Trotter J, Williams GM (1984) Can J Chem 62: 755
37. Takagi S, Miyamoto TK, Hamaguchi M, Sasaki Y, Matsumura T (1990) Inorg Chim Acta 173: 215
38. Collman JP, Barnes CE, Collins TJ, Brothers PJ, Galucci J, Ibers JA (1981) J Am Chem Soc 103: 7030
39. Massoudipour M, Pandey KK (1989) Inorg Chim Acta 160: 115
40. a) Eaton SS, Eaton GR, Holm RH (1972) J Organomet Chem 39: 179
    b) Eaton SS, Eaton GR (1975) J Am Chem Soc 97: 3660
41. Srivastava TS, Hoffman L, Tsutsui M (1972) J Am Chem Soc 94: 1385
42. Eaton SS, Eaton GR, Holm RH (1971) J Organomet Chem 32: C52
43. Sovocool GW, Hopf FR, Whitten DG (1972) J Am Chem Soc 94: 4350
44. Collman JP, Kim K, Leidner CR (1987) Inorg Chem 26: 1152
45. Franco C, McLendon G (1984) Inorg Chem 23: 2370
46. Buchler JW, Rohbock K (1974) J Organomet Chem 65: 223
47. Buchler JW, Folz M (1977) Z Naturforsch B32: 1439
48. Buchler JW, Herget G, Oesten K (1983) Liebigs Ann Chem p 2164
49. Collman JP, Barnes CE, Woo LK (1983) Proc Natl Acad Sci (USA) 80: 7684
50. Che CM, Poon CK, Chung WC, Gray HB (1985) Inorg Chem 24: 1277
51. Buchler JW, Künzel FM (1994) Z Anorg Allg Chem 620: 888
52. a) Buchler JW, Pfeifer S unpublished experiments;
    b) Pfeifer S (1988) Doctoral Dissertation, Technische Hochschule Darmstadt
53. a) Sadisavan N, Fleischer EB J (1968) Inorg Nucl Chem 30: 591
    b) Fleischer EB, Sadasivan (1967) JCS Chem Commun p 159
54. Ogoshi H, Setsune J, Omura T, Yoshida Z (1975) J Am Chem Soc 97: 6461
55. Yoshida Z, Ogoshi H, Omura T, Watanabe E, Kurosaki T (1972) Tetrahedron Lett p 1077
56. Ogoshi H, Omura T, Yoshida Z (1973) J Am Chem Soc 95: 1666
57. James BR, Stynes DV (1972) J Am Chem Soc 94: 6225
58. Boschi T, Licoccia S, Tagliatesta P (1987) Inorg Chim Acta 126: 157
59. Aoyama Y, Yoshida T, Sakurai K, Ogoshi H (1986) Organometallics 5: 168
60. Thackray DG, Ariel S, Leung TW, Menon K, James BR, Trotter J (1986) Can J Chem 64: 2440
61. Wayland BB, Ba S, Sherry AE (1991) J Am Chem Soc 113: 5305
62. Abeysekera AM, Grigg R, Trocha-Grimshaw J, Viswanatha V (1977) JCS Perkin I p 1395.
63. Adler AD, Longo FR, Kampas F, Kim J (1970) J Inorg Nucl Chem 32: 2443
64. Karle Hanson L, Gouterman M, Hanson JC (1973) J Am Chem Soc 95: 4822
65. Boschi T, Licoccia S, Tagliatesta P (1986) Inorg Chim Acta 119: 191
66. Boschi T, Licoccia, Tagliatesta P (1988) Inorg Chim Acta 143: 235
67. Boschi T, Licoccia, Paolesse R, Tagliatesta P (1989) Inorg Chim Acta 163: 135
68. Kalyanasundaram K (1984) Chem Phys Lett 104: 357
69. Kadish KM, Yao CL, Anderson JE, Cocolios P (1985) Inorg. Chem. 24: 4515

70. Massoudipour M, Tewari SM, Pandey KK (1989) Polyhedron 8: 1447.
71. Ogoshi H, Setsune J, Yoshida Z (1978) J Organomet Chem 159: 317
72. Sugimoto H, Ueda N, Mori M (1982) JCS Dalton Trans p 1611
73. Treibs A (1971) Das Leben und Wirken von Hans Fischer, Hans Fischer-Gesellschaft, München
74. Eisner U, Harding MJC (1964) J Chem Soc p 4089
75. Buchler JW, Puppe L (1974) Justus Liebigs Ann Chem p 1046
76. Buchler JW, Herget G (1987) Z Naturforsch 42b: 1003
77. Buchler JW, Dreher C, Herget G (1988) Liebigs Ann Chem p 43
78. Kielman-van Luyt ECM, Canters GW (1979) Spectrochim Acta A 35: 1089
79. Wang D, Jin C, Huang Z, He S (1987) Huaxue Shiji 9: 47
80. Lavallee DK, White A, Diaz A, Battioni JP, Mansuy D (1986) Tetrahedron Lett, p 3521
81. Inhoffen HH, Buchler JW, Thomas R (1969) Tetrahedron Lett, p 1141
82. Stolzenberg AM, Schussel LJ (1991) Inorg Chem 30: 3205
83. Lahiri GK, Summers JS, Stolzenberg AM (1991) Inorg Chem 30: 5049
84. Buchler JW, Eikelmann G, Puppe L, Rohbock K, Schneehage HH, Weck D (1971) Liebigs Ann Chem 745: 135
85. Buchler JW, Lay KL, Stoppa H (1980) Z. Naturforsch 35b: 433
86. (a) Macquet JP, Theophanides T (1973) Can J Chem 51: 219
(b) Berjot M, Bernard L, Macquet JP, Theophanides T (1975) J Raman Spectrosc 4: 3;
(c) Macquet JP, Millard MM, Theophanides T (1978) J Am Chem Soc 100: 4741
87. Vogler A, Kunkely H, Rethwisch B (1980) Inorg Chim Acta 46: 101
88. Dorough GD, Miller JR, Huennekens FM (1951) J Am Chem Soc 73: 4315.
89. Haurowitz F, Clar E, Hermann Z, Kittel H, Münzberg FK (1935) Chem Ber 68: 1795
90. Somaya O (1967) Doctoral Thesis, Technische Hochschule Braunschweig
91. More KM, Eaton SS, Eaton GR (1981) J Am Chem Soc 103: 1087
92. Brown TG, Hoffman BM (1980) Mol Phys 39: 1073
93. Konishi S, Hoshino M, Imamura MJ (1982) Phys Chem 86: 4888
94. Banci L (1985) Inorg Chem 24: 782
95. Banci L (1985) Inorg Chim Acta 101: 155
96. (a) Okoh JM, Bowles N, Krishnamurthy M (1984) Polyhedron 3: 1077
(b) Okoh JM, Krishnamurthy M (1986) JCS Dalton Trans, p 449
97. Rothemund R, Menotti AR (1948) J Am Chem Soc 70: 1808
98. Fleischer EB, Laszlo A (1969) Inorg Nucl Chem Lett 5: 373
99. Jamin ME, Iwamoto RT (1978) Inorg Chim Acta 27: 135
100. Buchler JW, Lay KL, Smith PD, Scheidt WR, Rupprecht GA, Kenny JA (1976) J Organomet Chem 110: 109
101. Vogel E, Jux N, Rodriguez-Val E, Lex J, Schmickler H (1990) Angew Chem 102: 1431, Angew Chem Int Ed Engl 29: 1387 and references therein
102. Vogel E, Köcher M, Schmickler H, Lex J (1986) Angew Chem 98: 262, Angew Chem Int Ed Engl 25: 257
103. Vogel E (1990) Pure Appl Chem 62: 557
104. Li ZY, Huang JS, Che CM, Chang CK (1992) Inorg Chem 31: 2670
105. Licoccia S (1995) Struct Bonding (Berlin) this volume
106. Leznoff CC, Lever ABP (eds) "Phthalocyanines. Properties and Applications", Vol. 1 (1989), Vol. 2 (1993). VCH, New York
107. Berezin BD (1981) Coordination compounds of porphyrins and phthalocyanines, Wiley, Chichester
108. Omiya S, Tsutsui M, Meyer EF, Jr, Bernal I, Cullen DL (1980) Inorg Chem 19: 134
109. Farrell NP, Murray AJ, Thornback JR, Dolphin DH, James BR (1978) Inorgan Chim Acta 28: L144
110. Schultz H, Lehmann H, Rein M, Hanack M (1991) Struct Bonding [Berlin] 74: 41
111. Kobel W, Hanack M (1986) Inorg Chem 25: 103
112. (a) Hanack M, Osio-Barcina J, Witke E, Pohmer J (1992) Synthesis p 211
(b) Hanack M, Knecht S, Witke E, Haisch P (1993) Synthetic Metals 55–57: 873
113. Hanack M, Vermehren P (1990) Inorg Chem 29: 134
114. Hanack M, Gül A, Subramanian LR (1992) Inorg Chem 31: 1542
115. Hanack M, personal communication
116. Ercolani C, personal communication; Capobianchi A, Paoletti AM, Pennesi G, Rossi G, Caminiti R, Ercolani C (1994) Inorg Chem 33: 4635

117. (a) Sievertsen S, Schlehahn H, Homborg H (1993) Z anorg allg Chem 619: 1064 (b) Homborg H (1993) personal communication ; (c) Sievertsen S, Schlehahn H, Homborg H (1994) Z Naturforsch B 49: 50
118. Muralidharan S, Ferraudi G, Schmatz K. (1982) Inorg Chem 21: 2961.
119. Münz X, Hanack M (1988) Chem Ber 121: 235
120. (a) Ostendorp G, Sievertsen H, Homborg H (1994) Z Anorg Allg Chem 620: 279 (b) Sievertsen H, Ostendorp G, Homborg H (1994) Z Anorg Allg Chem 620: 290
121. Hanack M, Haisch P, Lehmann H, Subramanin (1993) Synthesis p 387
122. Fu G, Fu Y, Jayaraj K, Lever ABP (1990) Inorg Chem 29: 4090
123. Ni YP, Fitzgerald JP, Carroll P, Wayland BB (1994) Inorg Chem 33: 2029
124. Geisser P, Schmid P, Scheffold R (1980) Chimia 34: 279
125. Liang Y, An X (1986) Huaxue Xuebao 44: 964
126. Hansen CB, Hoogers GJ, Drenth W (1993) J Mol Catal 79: 153
127. Szulbinski W, Lapkowski M (1986) Inorg Chim Acta 123: 127
128. Hambright P (1989) Inorg Chim Acta 157: 95
129. Buchler JW, Künzel FM, Mayer U, Nawra M (1994) Fresenius J Anal Chem 348: 371
130. Krishnamurthy M (1977) Inorg Chim Acta 25: 215
131. Hambright P, Langley R (1987) Inorg Chim Acta 137: 209
132. Ashley KR, Shyu SB, Leipoldt (1980) Inorg Chem 19: 1613
133. McLendon G, Bailey M (1979) Inorg Chem 18: 2120
134. Schmehl RH, Whitten DG (1981) J Phys Chem 85: 3473
135. Kalyanasundaram K, Neumann-Spallart M (1982) J Phys Chem 86: 5163
136. Pasternack RF, Brigandi RA, Abrams MJ, Willams AP, Gibbs EJ (1990) Inorg Chem 29: 4483
137. Blinova JA, Vasil-ev VV, Shagisultanova GA (1994) Russ J Inorg Chem 39: 253
138. Buchler JW, Katzenmeier H, Nawra M, unpublished results
139. Herrmann O, Mehdi SH, Corsini A (1978) Can J. Chem 56: 1084
140. Krishnamurthy M (1978) Inorg Chem 17: 2242
141. Okoh JM, Krishnamurthy M (1991) Inorg Chim Acta 189: 233
142. Abou-Gamra Z, Harriman A, Neta P (1986) J Chem Soc Faraday Trans II 82: 2337
143. Scheidt WR, Lee YJ (1987) Struct Bonding (Berlin) 64: 1
144. Bonnet JJ, Eaton SS, Eaton GR, Holm RH, Ibers JA (1973) J Am Chem Soc 95: 2141
145. Little RG, Ibers JA (1973) J Am Chem Soc 95: 8583
146. Collman JP, Barnes CE, Swepston PN, Ibers JA (1987) J Am Chem Soc 106: 3500
147. Masuda H, Taga T, Osaki K, Sugimoto H, Mori M, Ogoshi H (1982) J. Am Chem Soc 103: 2199
148. Masuda H, Taga T, Osaki K, Sugimoto H, Mori M, Ogoshi H (1982) Bull Chem Soc Jpn 55: 3887
149. James BR, Dolphin D, Leung TW, Einstein FWB, Willis AC (1984) Can J Chem 62: 1238
150. Collman JP, Barnes CE, Brothers PJ, Collins JT, Ozawa T, Gallucci JC, Ibers JS (1984) J Am Chem Soc 106: 5151
151. Ball RG, Domazetis G, Dolphin D, James BR, Trotter J (1981) Inorg Chem 20: 1556
152. Collman JP, Brauman JI, Fitzgerald JP, Hampton PD, Naruta Y, Sparapany JW, Ibers JA (1988) J Am Chem Soc 110: 3477
153. Seyler JW, Fanwick PE, Leidner CR (1990) Inorg Chem 29: 2021
154. Camenzind MJ, James BR, Dolphin D, Sparapany JW, Ibers JA (1988) Inorg Chem 27: 3054
155. Ke M, Rettig SJ, James BR, Dolphin D (1987) J Chem Soc Chem Commun p 1110
156. Groves JT, Han Y, Engen van D (1990) J Chem Soc Chem Commun 436
157. Ke M, Sishta C, James BR, Dolphin D, Sparapany JW, Ibers JA (1991) Inorg Chem 30: 4766
158. Seyler JW, Fanwick PE, Leidner CR (1992) Inorg Chem 31: 3699
159. James BR, Pacheco A, Rettig SJ, Ibers JA (1988) Inorg Chem 27: 2414
160. Cullen D, Meyer E, Srivastava TS, Tsutsui M (1972) J Chem Soc Chem Commun 584
161. Masuda H, Taga T, Osaki K, Sugimoto H, Mori M (1984) Bull Chem Soc Jpn 57: 2345
162. Che CM, Huang JS, Li ZY, Poon CK, Tong WF, Lai TF, Cheng MC, Wang CC, Wang Y (1992) Inorg Chem 31: 5220
163. Nasri H, Scheidt WR (1990) Acta Cryst C46: 1096
164. Che CM, Chung WC, Lai TF (1988) Inorg Chem 27: 2801
165. Smieja JA, Omberg KM, Breneman GL (1994) Inorg Chem 33: 614
166. Che CM, Lai TF, Chung WC, Schaefer WP, Gray HB (1987) Inorg Chem 26: 3907
167. Woo LK, Smith DA, Young VG (1991) Organometallics 10: 3977
168. Collman JP, Bohle DC, Powell AK (1993) Inorg Chem 32: 4004

169. Fleischer EB, Florian R (1973) Inorg Nucl Chem Lett 9: 1303
170. Grigg R, Trocha-Grimshaw J, Henrick K (1982) Acta Crystallogr Sect B B38: 2455
171. Takenaka A, Syal SK, Sasada Y, Omura T, Ogoshi H, Yoshida Z-I (1976) Acta Crystallogr Sect B B32: 62
172. Whang D, Kim K (1991) Acta Cryst C47: 2547
173. Takenaka A, Sasada Y, Ogoshi H, Omura T, Yoshida Z-I (1975) Acta Crystallogr Sect B B31, 1.
174. Jones NL, Carroll PJ, Wayland BB (1986) Organometallics 5: 33
175. Boschi T, Licoccia S, Paolesse R, Tagliatesta (1989) Organometallics 8: 330
176. Miller RG, Kyle JA, Coates GW, Anderson DJ, Fanwick PE (1993) Organometallics 12: 1161
177. Latos-Grazynski L, Lisowski J, Olmstead MM, Balch AL (1989) Inorg Chem 28: 3328
178. Cornillon JL, Anderson JE, Swistak C, Kadish KM (1986) J Am Chem Soc 108: 7633
179. Swistak C, Cornillon JL, Anderson JE, Kadish KM (1987) Organometallics 6: 2146
180. Kadish KM, Cornillon JL, Mitaine P, Deng YJ, Korp JD (1989) Inorg Chem 28: 2534
181. Kadish KM, Deng YJ, Korp JD (1990) Inorg Chem 29: 1036
182. Chan KS, Chen XM, Mak TCW (1992) Polyhedron 11: 2703
183. Fleischer EB, Miller CK, Webb LE (1964) J Am Chem Soc 86: 2342
184. Stolzenberg AM, Schussel LJ, Summers JS, Foxman BM, Petersen JL (1992) Inorg Chem 31: 1678
185. Golder AJ, Nolan KB, Povey DC (1988) Acta Cryst C44: 1916
186. Latos-Grazy'nski L, Lisowski J, Chmielewski P, Grzeszczuk M, Olmstead MM, Balch AL (1994) Inorg Chem 33: 192
187. Hazell A (1984) Acta Crystallogr C40: 751
188. Scheidt WR, Mondal JU, Eigenbrot CW, Adler A, Radonovich LJ, Hoard JL (1986) Inorg Chem 25: 795
189. Timkovich R, Tulinski A (1977) Inorg Chem 16: 962
190. Buchler JW, Kokisch W, Smith PD (1978) Struct. Bonding [Berlin] 34: 1
191. Buchler JW, Smith PD, unpublished; Smith PD (1976) Doctoral Dissertation, Technische Hochschule Aachen
192. Tsutsui M, Ostfeld D, Hoffman LM (1971) J Am Chem Soc 93: 1820
193. Ogoshi H, Sugimoto H, Yoshida Z (1978) Bull Chem Soc Jpn 51: 2369
194. Eaton SS, Eaton GR (1975) J Am Chem Soc 97: 235
195. Antipas A, Buchler JW, Gouterman M, Smith PD (1978) J Am Chem Soc 100: 3015
196. Buchler JW, Smith PD (1976) Chem Ber 109: 1465
197. Mosseri S, Neta P, Hambright P, Sabry DY, Harriman A (1988) J Chem Soc Dalton Trans 2705
198. Sishta C, Camenzind MJ, James BR, Dolphin D (1987) Inorg Chem 26: 1181
199. Eaton SS, Eaton GR (1975) Inorg Chem 15: 134
200. Gèze C, Legrand N, Bondon A, Simonneaux G (1992) Inorg Chim Acta 195: 73
201. Paulson DR, Bhakta SB, Hyun RY, Yuen M, Beaird CE, Lee SC, Kim I, Ybarra J (1988) Inorg Chim Acta 151: 149
202. Buchler JW, Kokisch W, unpublished; Kokisch W (1979) Doctoral Thesis, Technische Hochschule Aachen, 1979.
203. Buchler JW, Kokisch W (1981) Angew Chem 93: 418
204. Camenzind MJ, James BR, Dolphin D (1986) J Chem Soc J Chem Commun p 1137
205. Collman JP, Brauman JI, Fitzgerald JP, Sparapany JW, Ibers JA (1988) J Am Chem Soc 110: 3486
206. Buchler JW, Smith PD (1974) Angew Chem 13: 820
207. Buchler JW, Kokisch W , Smith PD, Tonn B (1978) Z Naturforsch 33b: 1371
208. Kadish KM, Hu Y, Tagliatesta P, Boschi T (1993) J Chem Soc Dalton Trans 1167
209. Rajapakse N, James BR, Dolphin D (1990) Can J Chem 68: 2274
210. Billecke J, Kokisch W, Buchler JW (1980) J Am Chem Soc 102: 3622
211. Dolphin D, James BR, Murray AJ, Thornback JR (1980) Can J Chem 58: 1125
212. Buchler JW (1978) Angew Chem 90: 425; Angew Chem Int Ed Engl 17: 407
213. Paulson DR, Addison AW, Dolphin D, James BR (1979) J Biol Chem 254: 7002
214. Srivastava TS (1977) Biochim Biophys Acta 491: 599
215. Gersonde K (1981) unpublished experiments; personal communication
216. Grund A, Kaltbeitzel A, Mathy A, Schwarz R, Bubeck C, Vermehren P, Hanack M (1992) J Phys Chem 96: 7450
217. Marvaud V, Launay JP (1993) Inorg Chem 32: 1376

218. Doeff MM, Sweigart DA (1981) Inorg Chem 20: 1683
219. Pomposo F, Carruthers D, Stynes DV (1982) Inorg Chem 21: 4245
220. Nyokong T, Guthrie-Strachan J (1993) Inorg Chim Acta 208: 239
221. Luk SY, Williams JO (1989) J Chem Soc Chem Commun p 158
222. Guilard R, Lecomte C, Kadish KM (1987) Struct Bonding [Berlin] 64: 205
223. Brothers PJ, Collman JP (1986) Acc Chem Res 19: 209
224. (a) Collman JP, Garner JM, Woo LK (1989) J Am Chem Soc 111: 8141
(b) Collman JP, Arnold HJ (1993) Accounts Chem Res 26: 586
225. Collman JP, Wagenknecht PS, Hutchinson JE (1994) Angew Chem 106: 1620. Angew Chem Int Ed Engl 33: 1537
226. Collman JP, Hutchison JE, Ennis MS, Lopez MA, Guilard R (1992) J Am Chem Soc 114: 8074
227. Collman JP, Hutchison JE, Lopez MA, Guilard R (1992) J Am Chem Soc 114: 8066
228. Collman JP, Hutchison JE, Lopez MA, Guilard R, Reed RA (1991) J Am Chem Soc 113: 2794
229. Collman JP, Ha Y, Wagenknecht PS, Lopez MA, Guilard R (1993 J Am Chem Soc 115: 9080
230. Collman JP, Wagenknecht PS, Hutchison JE, Nathan SL, Lopez MA, Guilard R, L'Her M, Bothner-By AA, Mishra PK (1992) J Am Chem Soc 114: 5654
231. Collman JP, Hutchison JE, Wagenknecht PS, Lewis NS, Lopez MA, Guilard R (1990) J Am Chem Soc 112: 8206
232. Collman JP, Garner JM (1990) J Am Chem Soc 112: 166
233. Collman JP, Arnold HJ, Fitzgerald JP, Weissman KJ (1993) J Am Chem Soc 115: 9309
234. Brown GM, Hopf FR, Meyer RJ, Whitten DG (1975) J Am Chem Soc 97: 5385
235. Malinski T, Chang D, Bottomley LA, Kadish KM (1982) Inorg Chem 21: 4248
236. Kadish KM, Chang D (1982) Inorg Chem 21: 3614
237. Rachlewicz K, Latos-Grazynski L (1988) Inorg Chim Acta 144: 213
238. Smith PD, Dolphin D, James BR (1981) J Organomet Chem 208: 239
239. Che CM, Chung WC (1986) J Chem Soc J Chem Commun p 386
240. Sugimoto H, Higashi T, Mori M, Nagano M, Yoshida Z, Ogoshi H (1982) Bull Chem Soc Jpn 55: 822
241. Buchler JW, Oesten K, unpublished; Oesten K (1981) Doctoral Thesis, Technische Hochschule Darmstadt
242. Sugimoto H, Mori M (1981) Chem Lett p 297
243. Groves JT, Quinn R (1984) Inorg Chem 23: 3844
244. Buchler JW, Smith PD (1974) Angew Chem Int Ed Engl 13: 341
245. Groves JT, Ahn KH (1987) Inorg Chem 26: 3831
246. Leung WH, Che CM (1989) J Am Chem Soc 111: 8812
247. Rachlewicz K, Grzeszczuk M, Latos-Grazynski L (1993) Polyhedron 12: 821
248. Huang JS, Che CM, Poon CK (1992) J Chem Soc J Chem Commun p 161
249. Huang JS, Che CM, Li Zy, Poon CK (1992) Inorg Chem 31: 1313
250. (a) Leung WH, Hun TSW, Wong KY, Wong WT (1994) J Chem Soc Dalton Trans 2713; (b) Smieja JA, Omberg KM, Busuego LN, Breneman GL (1994) Polyhedron 13: 339 ; (c) Che CM, Leung WH, Chung WC (1990) Inorg Chem 29: 1841
251. Sishta C, Ke M, James BR, Dolphin D (1986) J Chem Soc Chem Commun p 787
252. Serpone N, Jamieson MA, Netzel TL. (1981) J Photochem 15: 295
253. Greenhorn R, Jamieson MA, Serpone N (1984) J Photochem 27: 287
254. James BR, Mikkelsen SR, Leung TW, Williams GM, Wong R (1984) Inorg Chim Acta 85: 209
255. Che CM, Huang JS, Li ZY, Poon CK (1991) Inorg Chim Acta 190: 161
256. Tokita Y, Yamaguchi K, Watanabe Y, Morishima I (1993) Inorg Chem 32: 329
257. Buchler JW, Billecke J, Kokisch W (1980) In: Bannister JV, Hill HAO (eds) "Active Oxygen in Biological Species", Developments in Biochemistry, Elsevier, Amsterdam, Vol 11A, p 45
258. Paeng IR, Nakamoto K (1990) J Am Chem Soc 112: 3289
259. Lewandowski W, Proniewicz LM, Nakamoto K (1991) Inorgan Chim Acta 190: 145.
260. Collman JP, Brothers PJ, McElwee-White L, Rose E, Wright LJ (1985) J Am Chem Soc 107: 4570
261. Collman JP, Brothers PJ, McElwee-White L, Rose E (1985) J Am Chem Soc 107: 6110
262. Collman JP, Wagenknecht PS, Lewis NS (1992) J Am Chem Soc 114: 5665
263. Boschi T, Licoccia S, Paolese R, Tagliatesta P (1988) Inorgan Chim Acta 145: 19
264. Cohen IA, Chow BC (1974) Inorg Chem 13: 488
265. Hoshino M, Yasufuku K (1985) Chem Phys Lett 117: 259
266. Liu YH, Anderson JE, Kadish KM (1988) Inorg Chem 27: 2320
267. Thomas NC (1986) Transition Met Chem 11: 425

268. Sakurai H, Uchikubo H, Ishizu K, Tajima K, Aoyama Y, Ogoshi H (1988) Inorg Chem 27: 2691
269. Setsune J, Yoshida Z, Ogoshi H (1982) J Chem Soc Perkin Trans 1: 983
270. Collman JP, Kim K (1986) J Am Chem Soc 108: 7847
271. (a) Wayland BB, Sherry AE, Poszmik G, Bunn AG (1992) J Am Chem Soc 114: 1673
(b) Hoshino M, Yasufuku K, Seki H, Yamazaki H (1985) J Phys Chem 89: 3080
272. Del Rossi KJ, Wayland BB (1986) J Chem Soc Chem Commun p 1653
273. Wayland BB, Newman AR (1981) Inorg Chem 20: 3093
274. Sherry AE, Wayland BB (1989) J Am Chem Soc 111: 5010
275. (a) Hoshino M, Yasufuku K, Konishi S, Imamura M (1984) Inorg Chem 23: 1982
(b) Yamamoto S, Hoshino M, Yasufuku K, Imamura M (1984) Inorg Chem 23: 195
(c) Huang JW, Ji LN, Hsieh AK, Andy Hor TS (1992) Transition Met Chem 17: 280
276. Hoshino M, Yasufuku K (1985) Inorg Chem 24: 4408
277. Lee S, Mediati M, Wayland B (1994) J Chem Soc Chem Commun 2299
278. Del Rossi KJ, Wayland BB (1985) J Am Chem Soc 107: 7941
279. Collman JP, Ha YY, Guilard R, Lopez M-A (1993) Inorg Chem 32: 1788
280. Jones NL, Carroll PJ, Wayland BB (1986) Organometallics 5: 33
281. Bosch WH, Wayland BB (1986) J Organomet Chem 317: C5
282. Aoyama Y, Aoyagi K, Toi H, Ogoshi H (1983) Inorg Chem 22: 3046
283. Aoyama Y, Yamagishi A, Tanaka Y, Toi H, Ogoshi H (1987) J Am Chem Soc 109: 4735
284. Aoyama Y, Yamagishi A, Tanaka Y, Toi H, Ogoshi H (1988) J Am Chem Soc 110: 4076
285. Aoyama Y, Asakawa M, Yamagishi A, Toi H, Ogoshi H (1990) J Am Chem Soc 111: 3145
286. Aoyama Y, Nonaka S, Motomura T, Ogoshi H (1989) Chem Lett 1877
287. Aoyama Y, Uzawa T, Saita K, Tanaka Y, Toi H, Ogoshi H (1988) Tetrahedron Lett 29: 5271
288. Aoyama Y, Motomura T, Ogoshi H (1989) Angew Chem 101: 922
289. Ogoshi H, Hatakeyama H, Yamamura K, Kuroda Y (1990) Chem Lett 1990: 51
290. Hanack M, Münz X (1985) Synthetic Metals 10: 357
291. Münz X, Hanack M (1988) Chem Ber 121: 239
292. Felton RH (1978) In: Dolphin D (ed) The Porphyrins, Academic, New York, Vol. V, pp 53–126
293. Fuhrhop JH (1975) In: Smith KM (ed) Porphyrins and Metalloporphyrins, Elsevier, Amsterdam, pp 593–624
294. Kadish KM (1986) Progr Inorg Chem 34: 437
295. Antipas A, Gouterman M (1983) J Am Chem Soc 105: 4896
296. Buchler JW, Herget G, unpublished experiments; Herget G (1986) Doctoral Thesis, Technische Hochschule Darmstadt
297. Kim D, Holten D, Gouterman M, Buchler JW (1984) J Am Chem Soc 106: 4015
298. Stanley KD, Luo L, Lopez de la Vega R, Quirke JME (1993) Inorg Chem 32: 1233
299. Godziela GM, Goff H (1986) J Am Chem Soc 108: 2237
300. Okoh JM, Krishnamurthy M (1986) J Chem Soc Dalton Trans 449.
301. Krishnamurthy M, Sutter R (1985) Inorg Chem 24: 1943
302. Langley R, Hambright P, Williams RFX (1985) Inorgan Chim Acta 104: L25
303. Segawa H, Azumi R, Shimizu T (1992) J Am Chem Soc 114: 7564
304. McGhee EM, Hoffman BM, Ibers JA (1991) Inorg Chem 30: 2162
305. Zhong ZJ, Okawa H, Aoki R, Kida S (1988) Inorgan Chim Acta 144: 233
306. Guilard R, Kadish KM (1988) Chem. Rev. 88: 1121
307. Collman, JP, McElwee-White L, Brothers PJ, Rose E (1986) J Am Chem Soc 108: 1332
308. Seyler JW, Leidner CR (1990) Inorg Chem 29: 3636
309. Seyler JW, Safford LK, Leidner CR (1992) Inorg Chem 31: 4300
310. Seyler JW, Safford LK, Fanwick PE, Leidner CR (1992) Inorg Chem 31: 1545
311. Alexander CS, Rettig SJ, James BR (1994) Organometallics 13: 2542
312. Chan YW, Renner MW, Balch AL (1983) Organometallics 2: 1888
313. (a) Woo LK, Smith DA (1992) Organometallics 11: 2344
(b) Djukic JP, Smith DA, Young jr VG, Woo LK (1994) Organometallics 13: 3020
(c) Djukic JP, Young VG, Woo LK (1994) Organometallics 13: 3995
314. Chan YW, Wood FE, Renner MW, Hope H, Balch AL (1984) J Am Chem Soc 106: 3380
315. Balch AL, Chan YW, Olmstead MM, Renner MW, Wood FE (1988) J Am Chem Soc 110: 3897
316. Collman JP, Rose E, Venburg GD (1993) J Chem Soc Chem Commun 934
317. (a) Abeysekera AM, Grigg R, Trocha-Grimshaw J, Viswanatha V (1976) J Chem Soc Chem Commun 227
(b) Abeysekera AM, Grigg R, Trocha-Grimshaw J, Viswanatha V (1977) J Chem Soc Perkin Trans I: 36

318. (a) Ogoshi H, Setsune J, Yoshida Z (1975) J Chem Soc Chem Commun 572
(b) Ogoshi H, Setsune J, Yoshida Z (1980) J Organomet Chem 185: 95
319. James BR, Stynes, DV (1972) J Chem Soc Chem Commun 1261
320. (a) Callot HJ, Schaeffer E (1978) J Chem Soc Chem Commun 937
(b) Callot HJ, Schaeffer E (1980) Nouv J Chim 4: 311
321. Callot HJ, Metz F, Piechocki C (1982) Tetrahedron 38: 2365
322. (a) Maxwell JL, O'Malley S, Brown KC, Kodadek T (1992) Organometallics 11: 645
(b) O'Malley S, Kodadek T (1992) Organometallics 11: 2299
323. (a) Maxwell J, Kodadek T (1991) Organometallics 10: 4
(b) Bartley DW, Kodadek TJ (1993) J Am Chem Soc 115: 1656
324. Setsune J, Yazawa T, Ogoshi H, Yoshida Z (1980) J Chem Soc Perkin Trans I: 1641
325. Buchler JW, Puppe L, Schneehage HH (1971) Liebigs Ann Chem 135
326. Ogoshi H, Setsune JI, Nanbo Y, Yoshida ZI (1978) J Organomet Chem 159: 329
327. Aoyama Y, Yoshida T, Sakurai K, Ogoshi H (1983) J Chem Soc Chem Commun 478
328. Aoyama Y, Yoshida T, Ogoshi H (1985) Tetrahedron Lett 26: 6107
329. Wayland BB, Woods BA (1981) J Chem Soc Chem Commun 700
330. Wayland BB, Van Voorhees SL, Wilker C (1986) Inorg Chem 25: 4039
331. Wayland BB, Duttaahmed A, Woods BA (1983) J Chem Soc Chem Commun 142
332. Wayland BB, Woods BA, Pierce R (1982) J Am Chem Soc 104: 302
333. Farnos MD, Woods BA, Wayland BB (1986) J Am Chem Soc 108: 3659
334. Wayland BB, Woods BA, Minda VM (1982) J Chem Soc Chem Commun 634
335. van Voorhees SL, Wayland BB (1985) Organometallics 4: 1887
336. Bosch HW, Wayland BB (1986) J Chem Soc Chem Commun 900
337. Wayland BB, van Voorhes SL, Del Rossi KJ (1987) J Am Chem Soc 109: 6513
338. Paonessa RS, Thomas NC, Halpern J (1985) J Am Chem Soc 107: 4333
339. Wayland BB, Woods BA (1981) J Chem Soc Chem Commun 475
340. Wayland BB, Woods BA, Coffin VL (1986) Organometallics 5: 1059
341. Coffin VL, Brennen W, Wayland BB (1988) J Am Chem Soc 110: 6063
342. Bunn AG, Wayland BB (1992) J Am Chem Soc 114: 6917
343. Wayland BB, Feng Y, Ba S (1989) Organometallics 8: 1438
344. Wayland BB, Poszmik G, Fryd M (1992) Organometallics 11: 3534
345. Mizutani T, Uesaka T, Ogoshi H (1995) Organometallics 14: 341
346. Kadish KM, Mu XH (1990) Pure Appl Chem 62: 1051
347. Davis DG (1978) In: Dolphin D (ed) The Porphyrins, Academic, New York, Vol. V, pp 127–152
348. Collman JP, Prodolliet JW, Leidner CR (1986) J Am Chem Soc 108: 2916
349. Bettelheim A, Ozer D, Harth R, Murray RW (1988) J Electroanal Chem 246: 139
350. Mu XH, Kadish KM (1990) Langmuir 6: 51
351. Deng J, Mu XH, Kadish KM (1991) Inorg Chem 30: 1957
352. Anderson JE, Yao C-L, Kadish K (1986) Inorg Chem 25: 3224
353. Kadish KM, Hu Y, Boschi T, Tagliatesta P (1993) Inorg Chem 32: 2996
354. Yao C-L, Anderson JE, Kadish KM (1987) Inorg Chem 26: 2725
355. Nyokong T (1994) J Chem Soc Dalton Trans 1359
356. Kadish KM, Deng YJ, Yao C–L, Anderson JE (1988) Organometallics 7: 1979
357. Gisselbrecht JP, Gross M, Köcher M, Lausmann M, Vogel E (1990) J Am Chem Soc 112: 8618
358. Arnold DP, Heath GA (1993) J Am Chem Soc 115: 12197
359. Kadish KM, Lin XQ, Ding JQ, Wu YT, Araullo C (1986) Inorg Chem 25: 3236
360. Tung HC, Chooto P, Sawyer DT (1991) Langmuir 7: 1635
361. Montanari F, Casella L (eds) (1994) Metalloporphyrin Catalyzed Oxidations, Kluver Academic, Dordrecht
362. Groves J, Quinn R (1985) J Am Chem Soc 107: 5790
363. Leung WH, Che CM, Yeung CH, Poon CK (1993) Polyhedron 12: 2331
364. Cheng SYS, Rajapakse N, Rettig SJ, James BR (1994) J Chem Soc Chem Commun 2669
365. Groves JT, Ahn KH, Quinn R (1988) J Am Chem Soc 110: 4217
366. Scharbert B, Zeisberger E, Poůlůs E (1995) J Organomet Chem 493: 143
367. Ohtake H, Higuchi T, Hirobe M (1992) J Am Chem Soc 114: 10660 and preceding papers
368. Rajapakse N, James BR, Dolphin D (1989) Catalysis Lett 2: 219
369. (a) Tavarès M, Ramasseul R, Marchon JC, Bachet B, Brassy C, Mornon JP (1992) J Chem Soc Perkin Trans 2: 1321 and preceding papers
(b) Tavarès M, Ramasseul R, Marchon JC, Valleegoyet D, Gramain JC (1994) J Chem Research S: 74

370. (a) Le Maux P, Bahri H, Simonneaux G (1993) Tetrahedron 49: 1401
(b) Le Maux P, Bahri H, Simonneaux G (1991) J Chem Soc Chem Commun 1350
(c) LeMaux P, Bahri H, Simonneaux G (1994) J Chem Soc Chem Commun 1287
371. Licoccia S, Paci M, Tagliatesta P, Paolesse R, Anzonaroli S, Boschi T (1991) Magn Res in Chem 29: 1084
372. Aoyama Y, Tanaka Y, Yoshida T, Toi H, Ogoshi H (1987) J Organomet Chem 329: 251
373. Aoyama Y, Tanaka Y, Fujisawa T, Watanabe T, Toi H, Ogoshi H (1987) J Org Chem 52: 2555
374. Li X-M, Shinoda S, Saito Y (1989) J Mol Catal 49: 113
375. Pawlik M, Hoq MF, Shepherd RE (1983) J Chem Soc Chem Commun 1467
376. Bouwkampwijnoltz AL, Visscher W, Vanveen JAR (1994) Electrochim Acta 39: 1641
377. (a) Brunner H, Maiterth F, Treittinger B (1994) Chem Ber 127: 2141
(b) Obermeier H, Brunner H (1994) Angew Chem Int Ed Engl 33: 2214
378. Brun AM, Harriman A (1994) J Am Chem Soc 116: 10383
379. Gentemann S, Albaneze J, Garciaferrer R, Knapp S, Potenza JA, Schugar HJ, Holten D (1994) J Am Chem Soc 116: 281
380. McCleverty JA, Badiola JAN, Ward MD (1994) J Chem Soc Dalton Trans 2415.
381. Mackay LG, Anderson HL, Sanders JKM (1992) J Chem Soc Chem Commun 43
382. Drain CM, Lehn JM (1994) Chem Commun 2313

Metal Complexes of Corroles and Other Corrinoids

S. Licoccia and R. Paolesse

Department of Chemical Sciences and Technologies, University of Rome Tor Vergata, I-00173 Rome, Italy

Several tetrapyrroles are employed by Nature to carry out fundamental biological functions and synthetic models compounds have been extensively used in order to understand the mechanisms of action of natural systems.

One of the structures present in many enzymes is the corrinoid structure. It consists of a tetrapyrrolic macrocycle, where a direct link between two pyrrole rings exists. Such a direct link and the different degrees of unsaturation that can be introduced into the macrocycle modulate its chelating properties towards metal ions and its reactivity.

A most interesting example of the corrinoid structure is corrole, a macrocycle where an 18 electron aromatic π system analogous to that of a porphyrin is maintained. Corrole has been shown to be a versatile ligand capable of coordinating transition and main group metals without significant distortion of the macrocycle plane.

The present article reviews the developments of the chemistry of corrole and its metal complexes considering the synthetic procedures that can be followed in order to prepare such compounds, their spectroscopic characterization and redox reactivity and demonstrates the peculiar ligand field effect of this macrocycle.

The chemistry of other synthetic corrinoids is also reviewed with special attention to their preparation procedures and axial ligand binding reactions.

Structure and Bonding, Vol. 84

1 Introduction

Porphyrins and metalloporphyrins have been the subject of numerous investigations over the past century. This attention is not surprising if one considers that particular derivatives play central roles in essential biological processes such as photosynthesis, dioxygen transport and storage, and electron transfer.

From the point of view of the coordination chemist the porphyrinato ligand has turned out to be very versatile and almost all metals have been combined with a porphyrin. Such complexes have been used in a variety of applications ranging from mimicking biological functions to stereoselective catalysis.

The very rich chemistry of porphyrins has created a great deal of interest in other tetrapyrrolic macrocycles.

Despite such great interest, a question that is still unanswered is why modified tetrapyrrole ligands, like those found in factor F_{430} and vitamin B_{12}, are employed by natural systems to carry out specific chemistry rather than the porphyrin ligand. A possible explanation that has been proposed [1, 2] is that these modified tetrapyrrole ligands exhibit different flexibility as compared with porphyrins. Another very important factor, and probably the most important one in the case of corroles and corrinoids, is the difference in hole size between the various macrocycles. The tetrapyrrole that would most efficiently perform a specific function would be the one with the proper hole size for the radius of the metal ion involved in the process.

Despite the similarities existing among natural tetrapyrrolic macrocycles, e.g. they are all tetradentate equatorial N_4 ligands, structural variations cause a fine modulation of the reactivity of the central metal atom. Thus it is most probably the corrin skeleton that enables the cobalt atom of vitamin B_{12} to carry out reactions impossible for a similar cobalt porphyrinate.

Biologically essential tetrapyrrolic macrocycles of two main types exist: the porphyrins, the prosthetic group of heme proteins, and the corrins, most representative among which is the coenzyme of Vitamin B_{12}.

A close connection between the two groups has been established by the demonstration that the biogenetic route to them both involves, at an early stage, the pyrrole porphobilinogen [3]. Johnson and his group, in the 1960s, considered that such a species would generate a linear tetrapyrrole which evolution would lead to either the porphyrin or the corrin structure. With such biogenetic scheme in mind they approached the synthesis of these linear tetrapyrroles. They demonstrated [3, 4] that the same species, dihydrobilin (1,19-dideoxybiladiene-*a,c*) salts, can be used to prepare compounds containing either the porphyrin or the corrin skeleton.

It is now known [5] that the common intermediate to hemes, chlorophyll and vitamin B_{12} is uroporphyinogen III (Fig. 1). However, a very interesting outcome of Johnson's research is that a new tetrapyrrolic macrocycle: Corrole (Fig. 2) was synthesized for the first time.

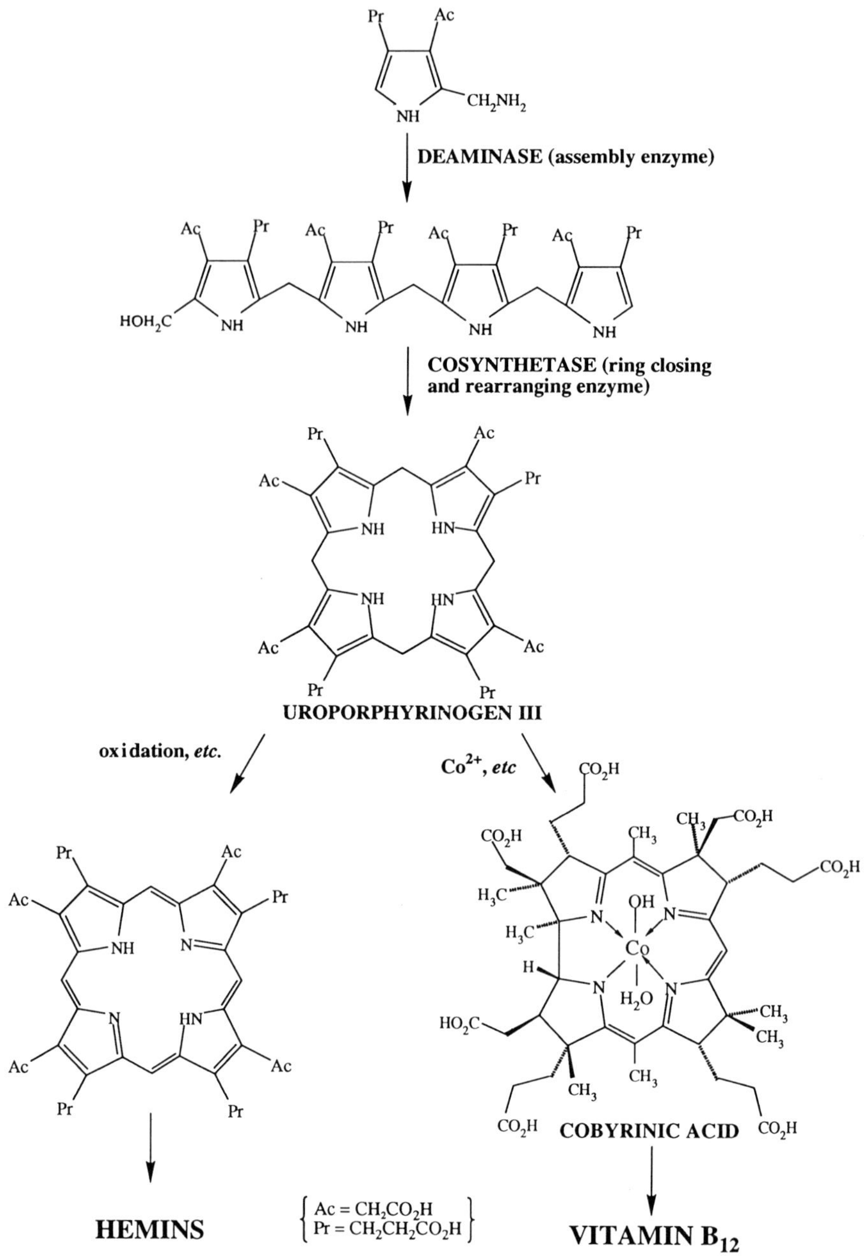

Fig. 1. Biosynthetic pathway to tetrapyrrolic macrocycles

Corrole is a tetrapyrrolic macrocycle with a direct link between two pyrrole rings. Lacking a C-20 *meso* carbon bridge it has a corrin like skeleton with double bonds involving porphyrin-like conjugation. It has an 18 electron π system and hence aromatic character. The direct link between the A and D

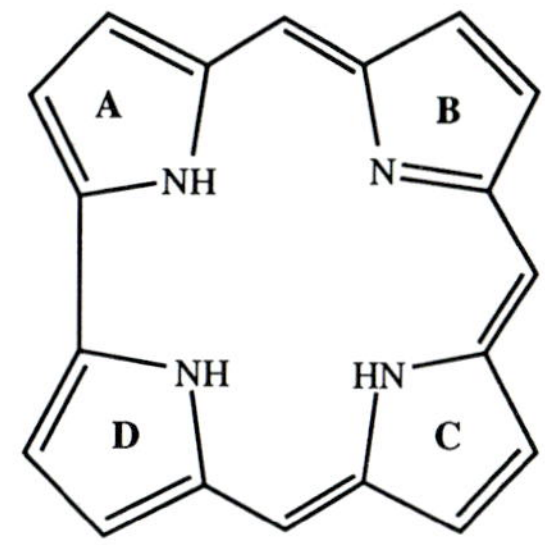

Fig. 2. The structural formula of corrole

pyrrole rings leads to short NH-NH contacts that cause a deformation of the macrocycle so that it cannot have a completely planar structure [6]. Such deviations from planarity are however quite small (8–10°) and the hybridization of carbon atoms at the 1- and 19-positions is considered to be sp^2. In one of its tautomeric forms corrole bears the same relation to porphyrins as does cyclopentadiene to benzene.

Another peculiarity of corrole is that, among tetradentate N_4 macrocyclic ligands, it completes the series corrin, porphyrin, corrole, where one, two or three amino-like nitrogens are present in the chelating system.

Corroles form an interesting class of compounds to be investigated in order to study how variations in symmetry and the degree of conjugation affect the chemistry of tetrapyrroles.

The chemistry of corroles and their metal complexes has been reviewed by different authors in the past [7–11] and apart from necessary clarifications this paper will deal with the contributions to the chemistry of metallocorrolates and corrinoids published in the last ten years.

With the name "corrinoid" we will refer to a class of compounds which have a molecular skeleton similar to that of the cobalt complex present in Vitamin B_{12} of which the main characteristics are the direct link between two pyrroles and the fully saturated β positions. These major features afford a macrocycle which is more contracted than a porphyrin and without an aromatic π system.

The definition of a synthetic route to coenzyme B_{12} has been a major concern of numerous groups. The synthesis was carried out several years ago, but the problem of the origin of the corrin structure in natural systems has not been resolved in all its steps [5, 12–14].

The characteristic reactions of the biological function of Vitamin B_{12}, such as methylations and isomerizations, are rather unusual in organic chemistry and numerous synthetic compounds have been proposed as models for the coenzyme.

In this review we will focus on synthetic derivatives of the macrocycles where a tetrapyrrolic skeleton is maintained and on their metal complexes without taking into account their biologically active counterparts.

2 Nomenclature

It is necessary to devote a specific section to the nomenclature of corrinoids and of their linear tetrapyrrolic precursors because of the confusion existing in the literature about the correct naming of these compounds.

An appropriate nomenclature has been recommended by the Commission for Biological Nomenclature of the International Union of Pure and Applied Chemistry and by the International Union of Biochemistry [15], but this approach has not been always accepted in the literature [16].

In order to avoid further elements of confusion both the IUPAC and traditional nomenclatures will be outlined here.

2.1 Linear Tetrapyrroles

The most commonly used approach for the naming of linear tetrapyrroles refers to a parent system named bilane, the structure of which is shown in Fig. 3a.

Positions can be numbered in two different manners but in most papers the *meso* positions have been designated with small italic letters in order to define the various unsaturations that can be introduced in these systems. Bilene-*b*, biladiene-*a,c*, bilatriene-*a,b,c*, for example, have been used to name the tetrapyrroles represented in Figs. 3b, 3c, 3d.

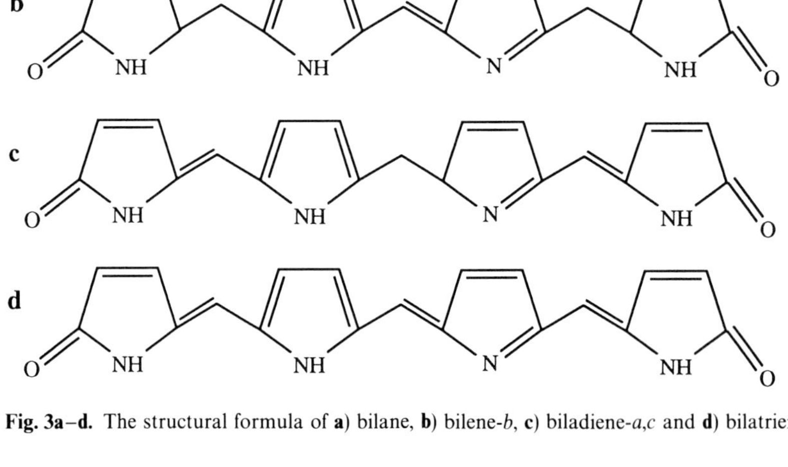

Fig. 3a–d. The structural formula of **a**) bilane, **b**) bilene-*b*, **c**) biladiene-*a,c* and **d**) bilatriene-*a,b,c*

The disadvantage of this approach is that it refers to a parent system containing oxygen atoms, so that for the linear tetrapyrroles without oxygen, which are those most commonly used, it is necessary to use the prefix 1,19-dideoxy [16].

IUPAC recommends the use of a parent system without oxygen and with the highest degree of unsaturation possible: this compound is named bilin and its formula is shown in Fig. 4.

This nomenclature procedure parallels that normally used for porphyrins, which is based on the compound with the highest level of unsaturation known: thus the porphyrin is the parent compound and the related reduced structures are named hydroporphyrins.

Following analogous considerations, reduced bilins are called hydrobilins. Positions are numbered as reported in Fig. 4, the number 20 is omitted in order to retain a numeration correlated to that of the porphyrin structure.

When we consider a proper linear precursor for a corrin, which is a highly reduced structure, a different parent tetrapyrrole may be used: this is named secocorrin and its shown in Fig. 5.

However, the name secocorrin has been used for compounds in different degrees of hydrogenation in the literature: in order to eliminate confusion it should therefore be avoided.

2.2 Corrinoids

The parent compound used for the nomenclature of corrinoids is the naturally occurring system: corrin, whose structure is shown in Fig. 6.

The ring is numbered as shown in the Figure and again the number 20 is omitted as in the case of linear tetrapyrroles; the pyrrole rings are designated with capital letters.

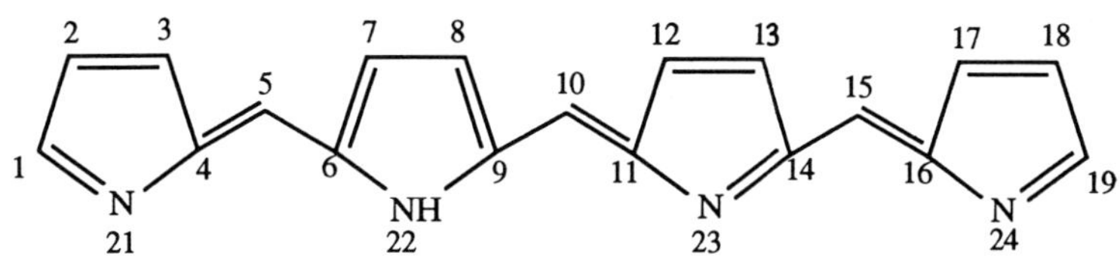

Fig. 4. The structural formula of bilin (22 H tautomer)

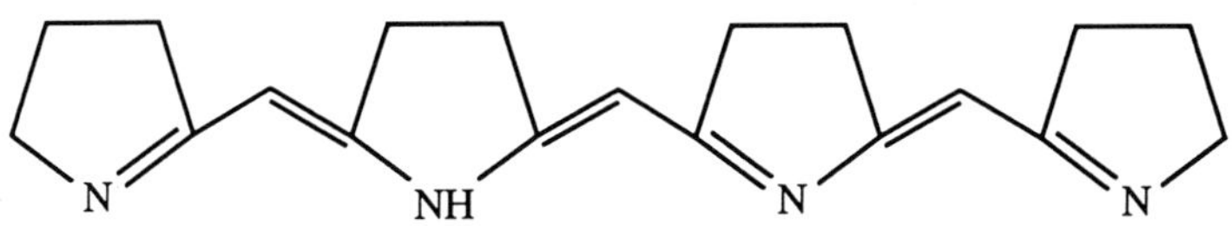

Fig. 5. The structural formula of secocorrin (1,2,3,7,8,12,13,17,18,19-decahydrobilin or 1,19-dideoxy-1,2,3,7,8,12,13,17,18,19-decahydrobilatriene-*a*,*b*,*c*)

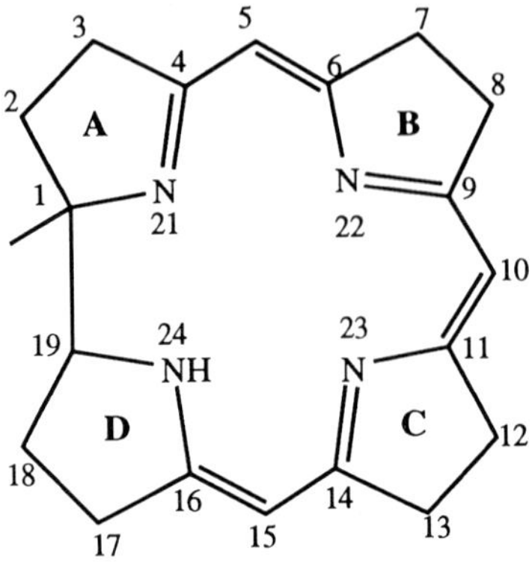

Fig. 6. The structural formula of corrin

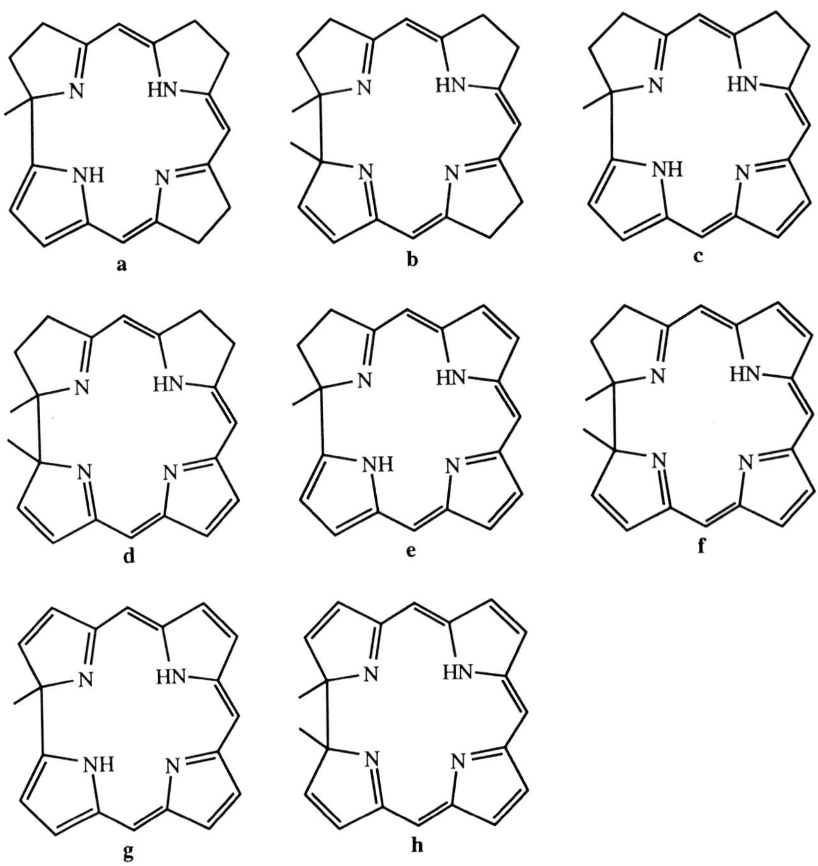

Fig. 7a–h. The structural formulae of: **a**) 1-methyl-D-didehydrocorrin, H_2(ADDC); **b**) 1,19-dimethyl-D-didehydrocorrin, H(A_2DDC); **c**) 1-methyl-C,D-tetradehydrocorrin, H_2(ATDC); **d**) 1,19-dimethyl-C,D-tetradehydrocorrin, H(A_2TDC); **e**) 1-methyl-B,C,D-hexadehydrocorrin, H_2(AXDC); (**f**) 1,19-dimethyl-B,C,D-hexadehydrocorrin, H(A_2XDC); **g**) 1-methyl-octadehydrocorrin, H_2(AODC); **h**) 1,19-dimethyl-octadehydrocorrin, H(A_2ODC)

At variance with what happens in the linear systems, the corrin represents the most saturated species: derivatives in which unsaturations are introduced, are named dehydrocorrins and the positions of the additional double bonds are referred to the pyrrole rings. In Fig. 7 several examples of such compounds are reported.

In the case of the most unsaturated species, such as corrole and octadehydrocorrin, some confusion arises since corrole is also formally an octadehydrocorrin, although with different hybridization of carbon atoms 1 and 19.

However, corrole (Fig. 2) represents a peculiar system and it is generally referred to with the non-systematic name which will also be used throughout this paper.

The name tetradehydrocorrin has also been used by the group of Johnson and Murakami for octadehydrocorrin: following the IUPAC recommendations the latter name will be used here in order to avoid confusion.

2.3 *Abbreviations*

In conclusions, the nomenclature used for tetrapyrroles and corrinoids that will be used in this paper is based on the species with the highest degree of unsaturation.

In most of the reported compounds β-positions are occupied by alkyl groups and only different substituent patterns will be explicitly indicated.

The following abbreviations will be used for the various macrocycles in their metal free, protonated neutral forms:

H(C)	Corrin
H_2(ADDC)	1-alkyl-didehydrocorrin
H(A_2DDC)	1,19-dialkyl-didehydrocorrin
H_2(ATDC)	1-alkyl-tetradehydrocorrin
H(A_2TDC)	1,19-dialkyl-tetradehydrocorrin
H_2(AXDC)	1-alkyl-hexadehydrocorrin
H(A_2XDC)	1,19-dialkyl-hexadehydrocorrin
H_2(AODC)	1-alkyl-octadehydrocorrin
H(A_2ODC)	1,19-dialkyl-octadehydrocorrin
H_2(CODC)	1-carboxyethyl-octadehydrocorrin
H(C_2ODC)	1,19-dicarboxyethyl-octadehydrocorrin
H(ACODC)	1-alkyl-19-carboxyethyl-octadehydrocorrin
H_3(OMC)	2,3,7,8,12,13,17,18-octamethylcorrole
H_3(TMTEC)	2,3,17,18-tetramethyl-7,8,12,13-tetraethylcorrole
H_3(OEC)	2,3,7,8,12,13,17,18-octaethylcorrole
H_3(OMMPC)	10-phenyl-2,3,7,8,12,13,17,18-octamethylcorrole
H_3(OMTPC)	5,10,15-triphenyl-2,3,7,8,12,13,17,18-octamethylcorrole
H_3(5,10-OMDPC)	5,10-diphenyl-2,3,7,8,12,13,17,18-octamethylcorrole
H_3(5,15-OMDPC)	5,15-diphenyl-2,3,7,8,12,13,17,18-octamethylcorrole

$H_3(TBC)$	triazatetrabenzcorrole
$H_2(P)$	porphyrin
$H_2(TPP)$	5,10,15,20-tetraphenylporphyrin
$H_2(OMP)$	2,3,7,8,12,13,17,18-octamethylporphyrin
H(Cby)	Cobyrinic acid (see Fig. 1)
H(Cbi)	Cobinamide: 1,5,7,12,12′,17-heptamethyl-2,7′,18-triacetamino-3,8,13-tripropionamino-17′-*N*-2-hydroxy-propionylamino-corrin (bis-cyano cobalt^{III} complex is shown in Fig. 31a)

Other abbreviations used:

Ac	$-CH_2CO_2H$
Pr	$-CH_2CH_2CO_2H$
AcO^-	$CH_3CO_2^-$
DMA	dimethylamine
Py	pyridine
DDQ	2,3-dichloro-5,6,-dicyano-1,4-benzoquinone

3 Syntheses of Metallocorrolates

Corrole (Fig. 2) is an aromatic tetradentate N_4 ligand in which three of four nitrogen atoms are linked to hydrogens that can be easily substituted. It readily forms stable anions which are also aromatic as demonstrated by their electronic spectra showing an intense absorption in the high energy region [8] (Fig. 8).

The synthetic routes leading to corroles have been reviewed in the past [10]. The main procedure involves the cyclization of dihydrobilins either via a photochemical process or a metal assisted one.

Corroles having an asymmetrical arrangement of the β-substituents have also been obtained by thermal cyclization of 1,19-dibromo or 1,19-diiodo-dihydrobilins.

Recently a major improvement of this latter procedure had been reported [17]. An efficient route for the preparation of both symmetrical and asymmetrical 1,19-dibromo-dihydrobilins has been developed. The cyclization of these open chain tetrapyrroles has been investigated under a variety of conditions and the best results for the preparation of corroles have been obtained in methanol.

Because of its structure, corrole could be expected to stabilize the +3 oxidation state for metal ions leading to the formation of neutral complexes.

Several metal ions have been reacted with corroles or *N*-alkylcorroles but, until 1980, the only example of a fully characterized metal^{3+} complex of corrole was a Co^{3+} derivative for which a crystal structure has been determined [11]. The formation of an Fe^{3+} derivative has also been reported but no detailed characterization of this complex has been carried out although it has been used in a catalytic application [18].

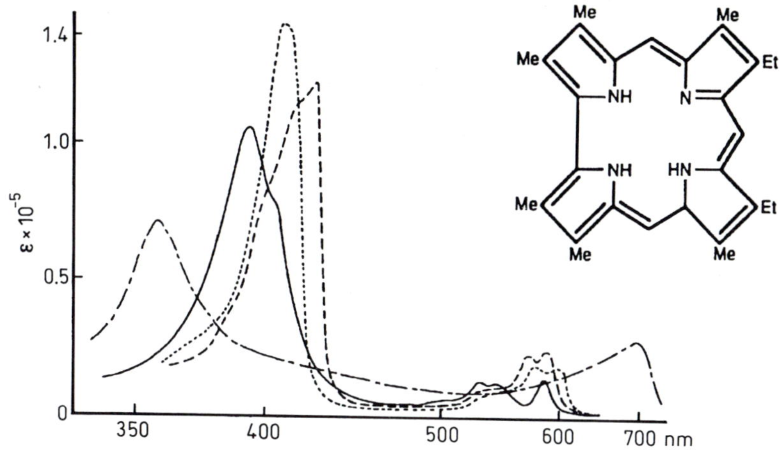

Fig. 8. Electronic spectra of neutral (——), anion (– – – – –), monocation (· · · · ·), and dication (· – · – · – ·) of a corrole. Taken from Ref. [8] with permission

																VII A	VIII A
1 H	2 He																2 He
3 Li	4 Be											5 B	6 C	7 N	8 O	9 F	10 Ne
11 Na	12 Mg											13 Al	14 Si	15 P	16 S	17 Cl	18 Ar
19 K	20 Ca	21 Sc	22 Ti	23 V	24* Cr	25* Mn	26* Fe	27* Co	28* Ni	29* Cu	30* Zn	31 Ga	32* Ge	33 As	34 Se	35 Br	36 Kr
37 Rb	38 Sr	39 Y	40 Zr	41 Nb	42* Mo	43 Tc	44 Ru	45* Rh	46* Pd	47 Ag	48 Cd	49* In	50* Sn	51 Sb	52 Te	53 I	54 Xe
55 Cs	56 Ba	57 La	72 Hf	73 Ta	74 W	75 Re	76 Os	77 Ir	78 Pt	79 Au	80 Hg	81 Tl	82 Pb	83 Bi	84 Po	85 At	86 Rn
87 Fr	88 Ra	89 Ac	104 Rf	105 Ha	106												

58 Ce	59 Pr	60 Nd	61 Pm	62 Sm	63 Eu	64 Gd	65 Tb	66 Dy	67 Ho	68 Er	69 Tm	70 Yb	71 Lu
90 Th	91 Pa	92 U	93 Np	94 Pu	95 Am	96 Cm	97 Bk	98 Cf	99 Es	100 Fm	101 Md	102 No	103 Lr

Fig. 9 The Periodic Table of Metallocorrolates. The *asterisk* marks the elements successfully introduced into a corrole ring

Thus, the historical development of the chemistry of metallocorrolates until 1980 includes complexes with Cu^{2+}, Ni^{2+}, Pd^{2+}, Fe^{3+}, Co^{3+}, Rh^{+}, Mo^{5+} and Cr^{5+}. The palladium complex has been isolated as its pyridinium salt since the neutral species was too unstable to be isolated or spectroscopically characterized [19]. The nickel complex was non-aromatic, with one of the potentially tautomeric hydrogens displaced from nitrogen to carbon in such a way as to interrupt the chromophore. In contrast the electronic spectrum of the paramagnetic copper complex is similar to those of the fully conjugated *N*(21)-methyl derivatives [11].

In the last ten years, several metal ions have been inserted into the corrole moiety leading to the Periodic Table of Metallocorrolates shown in Fig. 9.

The ligand used in most studies has been the totally symmetric octamethyl derivative $H_3(OMC)$.

Two major strategies have been considered in order to obtain metallocorrolates: the first one consists of the reaction of the preformed macrocycle with metal ions, while the second involves the oxidative, base induced cyclization of the open-chain tetrapyrrolic precursor dihydrobilin (1,19-dideoxybiladiene-*a,c*) in buffered alcoholic solution in the presence of metal salts. By proper tuning of the experimental conditions (reaction time, solvent, metal carrier) several metal complexes have been obtained and characterized.

3.1 Synthetic Methods Involving Reactions of the Preformed Macrocycle

The first synthetic method which involves the preformed corrole ring is the application to corroles of Adler's synthesis of metalloporphyrinates [20].

H_3OMC reacts with rhodium trichloride in *N,N*-dimethylformamide causing decarbonylation of the solvent and incorporation of one molecule of dimethylamine in the resulting metal complex Rh(OMC)DMA [21], following a pathway already observed in the case of porphyrins [22, 23].

The use of $RuCl_3$ also causes decomposition of the solvent. With such a metal, however, the product of the reaction is $H_2(OMP)$ [24]. The carbon monoxide derived from the decomposition of the solvent is added to the macrocycle in a ring expansion reaction generating the porphyrin ring.

Mn^{3+}, Fe^{3+} and In^{3+} complexes of H_3OMC can also be prepared by this method [24, 25].

Another procedure is the reaction of corrole with metal carbonyls in non-coordinating solvents such as toluene or benzene. Thus $Mn_2(CO)_{10}$, $Fe(CO)_5$ and $[Rh(CO)_2Cl]_2$ lead to the formation of the corresponding $metal^{3+}$ complexes. Also in this method the presence of an axial ligand is essential for the isolation of Rh^{3+} corrolates [21, 24].

If $Ru_3(CO)_{12}$ is used, the product is again the porphyrin, in a mixture with its Ru^{2+} complex. The catalytic action of ruthenium in expanding the macrocycle core has not been further investigated. The yield of ruthenium porphyrinate increases with time: this is not surprising since the reaction of triruthenium dodecacarbonyl with porphyrins is a standard procedure for the synthesis of such complexes [26].

Zn^{2+} corrolate can be obtained, as pyridinium salt, by reaction of corrole with zinc acetate in pyridine [25] in a procedure similar to that reported for the preparation of nickel and palladium complexes of corrole [11]. The zinc derivative is not paramagnetic and its formulation has been made on the basis of its proton NMR spectrum. Attempts to isolate the neutral zinc complex have been unsuccessful.

Iron corrolates can also be prepared reacting the macrocycle with metal carbonyls. The nature of the resulting metal complexes is however strongly dependent on the substituent pattern at the macrocycle periphery.

When H_3OMC is used, the complex isolated is the neutral Fe^{3+} complex Fe(OMC) [24] which has recently been thoroughly spectroscopically characterized (see Sects. 4.3, 4.5). The reaction of H_3OEC with iron carriers and the reactivity of the resulting metal complexes has been extensively investigated by Vogel et al. [27]. When H_3OEC is reacted with nonacarbonyldiiron in toluene under argon a brown species is formed. Admission of air causes change in the color of the solution from brown to red and yields μ-oxo-bis(octaethylcorrolato)iron (IV) in good yield [27]. The isolation of the Fe^{3+} complex in the reaction of H_3OMC has been ascribed to the poor solubility of this species that prevents the observation of the Fe^{3+}/Fe^{4+} oxidation.

Cleavage with HCl of the μ-oxo- complex yields the mononuclear Fe^{4+} complex Fe(OEC)Cl. The reaction is reversed upon addition of bases.

Reaction of Fe(OEC)Cl with phenylmagnesium bromide yields the σ-phenyl complex Fe(OEC)Ph, the first example of an organometallic corrolate [27].

Pentacoordinate Fe^{3+} corrolates bearing pyridine as axial ligand can be prepared by dissolving Fe(OMC) in pyridine [28], by metalation of H_3OEC with $Fe_2(CO)_9$ in the presence of pyridine in anaerobic conditions or by reduction of $[Fe(OEC)]_2O$ with hydrazine in the presence of pyridine [27]. Both complexes have been characterized spectroscopically (see Sect. 4.5) as Fe^{3+} derivatives.

In the case of Fe(OEC)Py susceptibility measurements indicated a $S = 3/2$ ground state for the metal atom ($\mu = 3.80$ BM) but this characterization is not in agreement with Mössbauer data ($\delta_{Fe} = -0.09\ mm\,s^{-1}$, $\Delta E_q = 3.88\ mm\,s^{-1}$) which correspond rather to those observed for the Fe^{4+} complexes. The hypothesis has been considered that the complex would be an Fe^{4+} corrole anion radical the magnetic properties ($S = 3/2$) of which would result from ferromagnetic spin coupling of Fe^{4+} ($S = 1$) with a macrocycle anion radical ($S = 1/2$) [27]. Theoretical studies are necessary to completely characterize the electron distribution in the complex.

All such synthetic procedures are summarized in Table 1.

Table 1. Metallocorrolates prepared by reaction of the metal free macrocycle with different metal carriers

Metal carrier	Macrocycle	Solvent	Complex	Reference
$Zn(AcO)_2$	H_3OMC	Pyridine	$[Zn(OMC)]^-[PyH]^+$	25
$Mn_2(CO)_{10}$	H_3OMC	Toluene	Mn(OMC)	24
$Fe(CO)_5$	H_3OMC	Toluene	Fe(OMC)	24
$RhCl_3$	H_3OMC	DMF	Rh(OMC)DMA	29
$Mn(AcO)_3$	H_3OMC	DMF	Mn(OMC)	24
$FeCl_3$	H_3OMC	DMF	Fe(OMC)	24
$InCl_3$	H_3OMC	DMF	In(OMC)	25
$Fe_2(CO)_9$	H_3OEC	Toluene	$[Fe(OEC)]_2O$	27
$Fe_2(CO)_9$	H_3OEC	Toluene/Py	Fe(OEC)Py	27

3.2 Synthetic Methods Involving Cyclization of Dihydrobilins

The reactions of dihydrobilin (1,19-dideoxybiladiene-*a*, *c*) with transition metals are strongly influenced by the nature of the metal ion. Thus with $Mn(OAc)_3$ or $FeCl_3$ the corresponding metallocorrolates have been obtained in high yield, in the presence of chromium or ruthenium salts the reaction product isolated has been the metal free macrocycle, while coordination of rhodium requires the presence of an axial ligand such as a phosphine, arsine or amine [21]. Neutral pentacoordinated rhodium complexes have thus been obtained. Although analysis of the electronic spectra of the reaction mixtures demonstrated that cyclization of the open-chain precursor and formation of metallocorrolates occur even in the absence of extra ligands, no axially unsubstituted rhodium derivative has been reported.

The stabilizing effect of an axial ligand has been previously observed in the synthesis of cobalt corrolates. Such an effect has been used to synthesize the complex where no peripheral β substituents are present on the macrocycle, which decomposes if attempts are made to isolate it in the absence of triphenylphosphine [10]. The behavior of rhodium closely resembled that of cobalt and it seems to be even more sensitive to the presence of axial ligands. $[Rh(CO)_2Cl]_2$ has also used as a metal carrier: with such a starting material a hexacoordinated derivative has been isolated. The reaction follows a pathway similar to that observed for rhodium porphyrinates: the first product is a Rh^+ complex which is then oxidized to a Rh^{3+} derivative [29].

Pentacoordinated Rh^{3+} corrolates react, in mild conditions, with isocyanides leading to the formation of hexacoordinated complexes $Rh(OMC)(CNR)_2$ [24].

The mechanism of cyclization has been investigated several times in the past [8]. It is catalyzed by the presence of a base: the first step of the reaction is, in fact, the formation of a green dihydrobilin free base. Visible light can then furnish the necessary energy for cyclization. Otherwise, once the dihydrobilin free base is formed, coordination of a metal ion achieves the correct geometry for cyclization. The metal exerts a template action in holding the reactive sites in proximity. The participation of the metal in the cyclization process is demonstrated by the fact that metal free corrole has been isolated as reaction product when Cr or Ru salts have been used, but no light was needed for the reaction to occur [24].

The same synthetic procedure has been applied to the preparation of main group metal derivatives of corrole [25]. Thus Sn(OMC)X, Ge(OMC)X and In(OMC) (where X is chloride or acetate) have been prepared starting from $SnCl_2$, $GeCl_4$ and $InCl_3$.

The nature of the counter ion in tin and germanium corrolates has been related to the base used in the cyclization: chloride complexes have been obtained using Et_3N, while acetate derivatives are the products when sodium acetate is used. Methatetical exchange of the counterions can be easily achieved.

Table 2 lists the metallocorrolates obtained by this synthetic procedure.

Table 2. Metallocorrolates prepared by reaction of 2,3,7,8,-12,13,17,18-octamethyl-dihydrobilin (1,19-dideoxy-2,3,7,8,12,-13,17,18-octamethyl-biladiene-*a,c*) with MX_n salts in methanolic solution in the presence of a base

MX_n	Base	Complex	Reference
$Mn(AcO)_3$	AcO^-	Mn(OMC)	24
$FeCl_3$	AcO^-	Fe(OMC)	24
$RhCl_3$	AcO^-	Rh(OMC)PPh_3	21
$RhCl_3$	AcO^-	Rh(OMC)PPh_2Me	21
$RhCl_3$	AcO^-	Rh(OMC)$AsPh_3$	21
$RhCl_3$	Et_3N	Rh(OMC)NEt_3	21
$[Rh(CO)_2Cl]_2$	AcO^-	RhCO(OMC)PPh_3	21
$SnCl_2$	Et_3N	Sn(OMC)Cl	25
$SnCl_2$	AcO^-	Sn(OMC)OA_c	25
$GeCl_4$	Et_3N	Ge(OMC)Cl	25
$GeCl_4$	AcO^-	Ge(OMC)OAc	25
$InCl_3$	AcO^-	In(OMC)	25

3.3 *Syntheses of* meso-*Substituted Corrolates*

Two papers have been published to date on *meso*-substituted derivatives of corrole [30, 31]. Four new cobalt complexes have been prepared and characterized: [triphenylphosphine(2,3,7,8,12,13,17,18-octamethyl-10-phenyl-corrolato)-cobalt (III)] (Co(OMMPC)PPh_3), [triphenylphosphine (2,3,7,8,12,13,17,18-octamethyl-5,10,15-triphenylcorrolato)cobalt (III)] (Co(OMTPC)PPh_3), [triphenylphosphine(2,3,7,8,12,13,17,18-octamethyl-5,10-diphenylcorrolato)cobalt (III)] (Co(5,10-OMDPC)PPh_3) and [triphenylphosphine (2,3,7,8,12,13,17,18-octamethyl-5,15-diphenylcorrolato)cobalt(III)] (Co(5,15-OMDPC)PPh_3).

The number and the position of the phenyl groups inserted at the periphery of the macrocycle have been the key factors in determining the synthetic procedure that has been designed in order to obtain the products.

The 10-substituted complex has been prepared following the procedure outlined in Sect. 3.2. i.e. reacting the appropriate tetrapyrrolic precursor, 2,3,7,8,12,13,17,18-octamethyl-10-phenyl-dihydrobilin, with cobalt acetate in methanol. As in the case of the *meso*-unsubstituted complex, the presence of triphenylphosphine is reported to be essential for the isolation of the product.

Four different synthetic procedures have been examined for the preparation of the triphenyl derivative, the fourth one suggested by the synthetic conditions developed to obtain the diphenyl derivatives. In the first three procedures it has been impossible to isolate the triphenyl-dihydrobilin. Its formation has been demonstrated, however, by monitoring the electronic spectrum of the reaction mixture and the cyclization to corrole has been carried out in situ. The synthesis that gave the highest yield (20%) and that avoids tedious purification procedures is outlined in Fig. 10. It involved the acidic condensation of benzaldehyde with two equivalents of 3,3′,4,4′-tetramethyl-*meso*-phenyl-dipyrromethane-5,5′

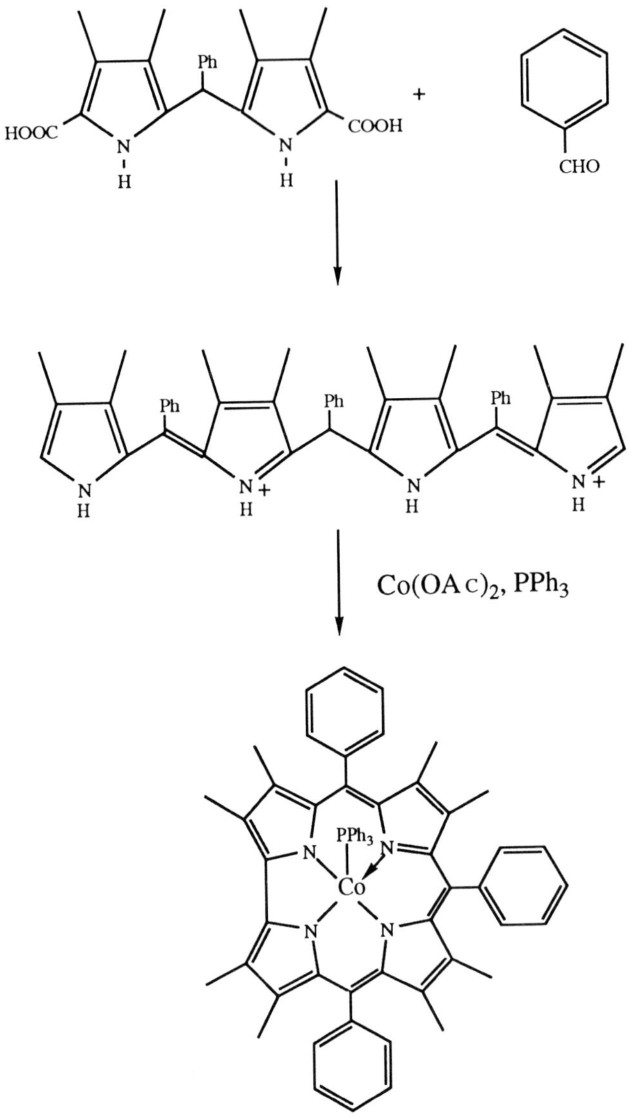

Fig. 10. Synthetic scheme leading to Co(OMTPC)PPh_3. According to Ref. [30]

dicarboxylic acid. The reaction mixture was then basified by the addition of triethylamine and reacted with $Co(OAc)_2$ in the presence of triphenylphosphine. The addition of DDQ allowed the required oxidative step.

More recently [31] the same authors reported the synthesis of Co(OMTPC)PPh_3 via self-condensation of 3,4-dimethyl-2-(α-hydroxybenzyl)-pyrrole-5-carboxylic acid. The procedure represents the first example of direct formation of the corrole ring from a monopyrrolic precursor.

The cobalt atom is essential to drive the reaction towards the formation of the corrole ring: other metals, such as Mn, Fe or Rh, in similar conditions give the expected *meso*-tetraphenyl-octamethylporphyrin in a mixture with its metal complexes.

This unusual feature of the cobalt atom has been attributed to the steric relief allowed by the formation of the corrole ring. Co(OMTPC)PPh_3 has in fact a planar structure, as confirmed by its spectral properties in solution and by a single crystal X-ray analysis (see Sect. 4.1, 4.5) while analogous (*meso*-tetraphenyl-octaalkylporphyrinato)zinc complexes suffer significant deviations from planarity, due to the steric crowing at the peripheral positions. This hypothesis is supported by the observation that H_2TPP is the product of the reaction of 2-(α-hydroxybenzyl)pyrrole with cobalt salts in similar conditions: in this case, the absence of significant steric interactions allows the formation of the porphyrin ring.

The synthesis of cobalt *meso*-diphenyl corrolates has also been reported [31]. The synthetic procedure involves the acidic condensation of 3,4-dimethyl-2-(α-hydroxybenzyl)pyrrole-5-carboxylic acid with 3,3′,4,4′-tetramethyl dipyrromethane, followed by reaction with cobalt salts. The reaction afforded a mixture of two isomers: Co(5,15-OMDPC)PPh_3 and Co(5,10-OMDPC)PPh_3. The formation of this latter isomer has been explained by the high tendency of self condensation of the starting pyrrole: under the reaction conditions, 2-(α-hydroxybenzyl)*meso*-phenyl dipyrromethane can be formed. This species would afford the Co(5,10-OMDPC)PPh_3 by further condensation with the dipyrromethane unit present in excess in the reaction mixture.

4 Structural and Spectral Properties

4.1 Solid State Structures

The crystal structures reported for metallocorrolates in the last ten years are those of Rh(OMC)$AsPh_3$ [24], Co (OMTPC)PPh_3 [31] and those of several iron complexes: [Fe(OEC)]O, Fe(OEC)Cl, Fe(OEC)Ph and Fe(OEC)Py [27].

The structure of Rh(OMC)$AsPh_3$ was determined by X-ray crystallography as shown in Fig. 11. It consists of discrete monomeric units where the rhodium atom has a distorted square pyramidal geometry.

The basal plane is defined by the four nitrogen atoms of the corrole moiety and the apex of the pyramid is occupied by the arsenic atom.

The distortion from regular pyramidal geometry is mainly reflected in the two angles subtended at the metal by the diagonally opposite pair of nitrogen atoms which are significantly narrower than 180°. The four Rh-N bonds are not equivalent, those involving the A and D rings, the two pyrroles directly linked, being shorter (see Table 3).

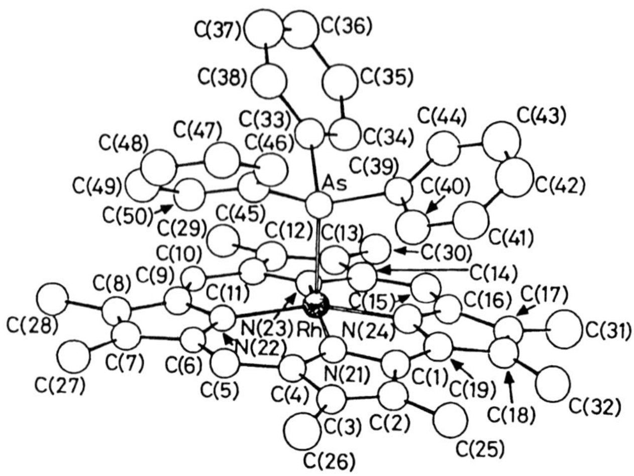

Fig. 11. A perspective view of $Rh(OMC)AsPh_3$ with the atom labeling scheme. Thermal ellipsoids are drawn at the 40% probability level. Taken from Ref. [24] with permission

Table 3. Selected bond angles and distances for $Rh(OMC)AsPh_3$ and $Co(OMTPC)PPh_3$. According to Refs. [22, 31]

N(21)-Rh-N(23)	163.2°
N(22)-Rh-N(24)	163.3°
Rh-N(21)	1.930 Å
Rh-N(24)	1.932 Å
Rh-N(22)	1.960 Å
Rh-N(23)	1.956 Å
N(21)-Co-N(23)	161.0°
N(22)-Co-N(24)	161.5°
Co-N(21)	1.879 Å
Co-N(24)	1.878 Å
Co-N(22)	1.900 Å
Co-N(23)	1.894 Å

The distortion is the result of the strain imposed on the whole molecule by the geometric requirements of the corrole structure. It is not, however, very significant: the rhodium atom is displaced by only 0.26 Å from the plane of the four coordinating nitrogen atoms, a value much smaller than that observed in the structure of the β-unsubstituted complex $Co(Corrole)PPh_3$ shown in Fig. 12 where the cobalt atom is displaced by 0.38 Å from the macrocycle plane [32]. In both compounds the four coordinating nitrogen atoms are strictly coplanar.

A most peculiar feature of the structure of $Rh(OMC)AsPh_3$ is that the 15 atom core of the corrole moiety has an almost planar conformation. The deviation from the plane of best fit is 0.1 Å and the maximum deviation of the rhodium atom from such mean plane is 0.19 Å.

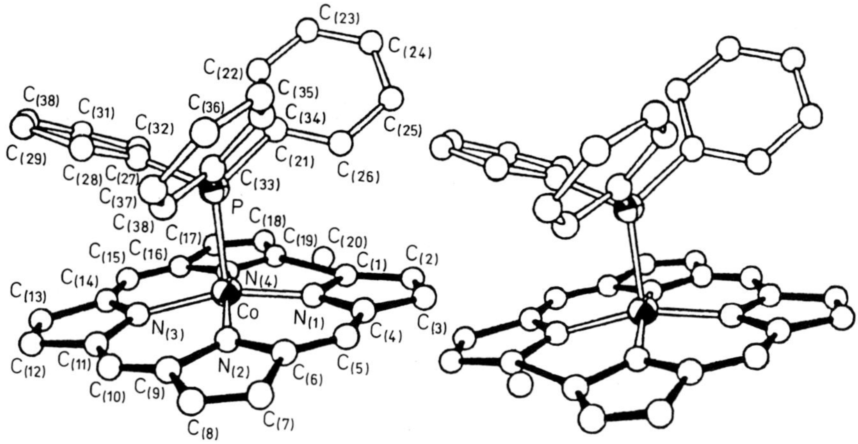

Fig. 12. The crystal structure of CoC Corrole)PPh_3. The Figure shows the two major orientations of the disordered corrole ring. Taken from Ref. [32] with permission

It is interesting to note that in the structure of both rhodium and cobalt complexes the corrole framework shows a higher degree of planarity than that observed in the free ligand where a large distortion is induced by contacts between the inner hydrogen atoms [11]. The free base corrole macrocycle in its unprotonated form seems to be very flexible and the macrocycle core can expand to accommodate a large metal ion such as Rh^{3+} in a *quasi*-planar structure.

Figure 13a illustrates the geometry of the complex Co(OMTPC)PPh_3 [31]. As in the case of the Rh^{3+} complex the tetraazamacrocyclic corrole moiety acts as a trianionic ligand to the metal atom which resides in an approximately square pyramidal coordination environment. The nitrogen atoms occupy the four coordination sites of the basal plane, while the apex of the pyramid is occupied by the phosphorus atom. The steric requirements of the phosphine group severely distort the geometry of the complex and result in non-orthogonal angles at the cobalt atom which is displaced 0.28 Å out of the equatorial plane towards the phosphorus, a value comparable with that found in Rh(OMC)$AsPh_3$. The structures of the two complexes exhibit close geometrical resemblance and are roughly superimposable as shown in Fig. 13b. Selected data are reported in Table 3.

The crystal structure of $[Fe(OEC)]_2O$ is shown in Fig. 14a. The metal atoms are coordinated in an approximately square pyramidal geometry. The corrole ligands have ring frameworks that are twisted to varying degrees because of steric interactions between the ethyl groups of the upper and lower part of the molecule. Although it would be expected to be linear on the basis of theoretical arguments, the Fe–O–Fe unit exhibits some twisting, the Fe–O–Fe angle being 170°.

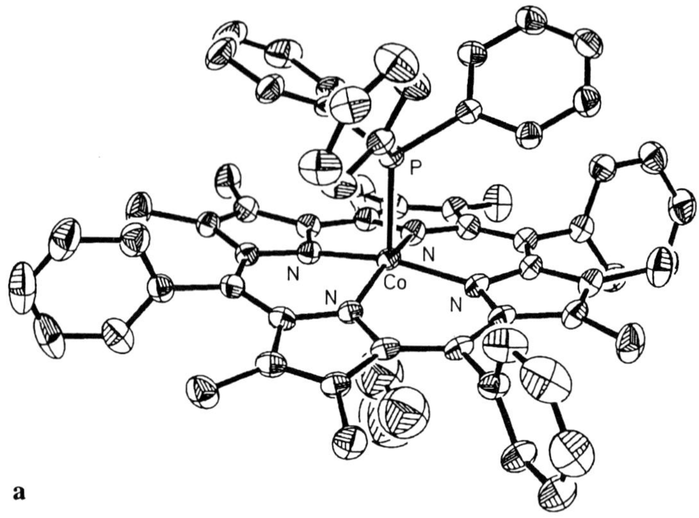

a

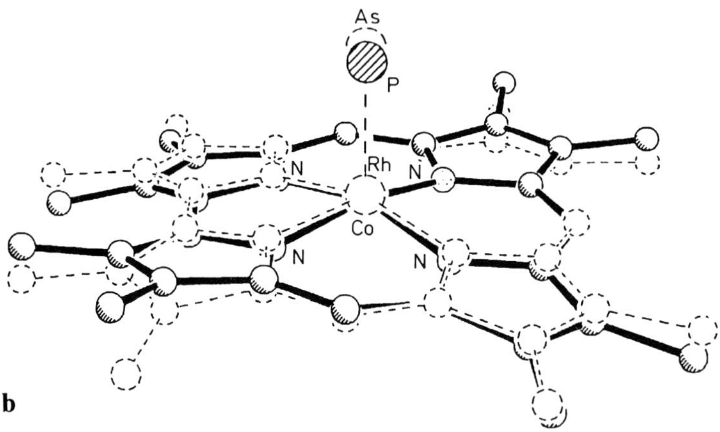

b

Fig. 13. a) The crystal structure of Co(OMTPC)PPh_3. Taken from Ref. [31] with permission. **b**) Superimposition of the structures of Co(OMTPC)PPh_3 and Rh(OMC)$AsPh_3$

The coordinative environment of the metal atom in the structures of Fe(OEC)Cl (Fig. 14b) and Fe(OEC)Ph (Fig. 14c) is similar to that found in the μ-oxo complex but in both complexes the ligand assumes a more planar conformation. The σ-bonded complex exists in the crystal as a π-π dimer, the intermolecular distance being 3.5–3.6 Å.

In the pyridine complex (Fig. 14d) the deviation from planarity is larger than that observed in Fe(OEC)Ph, the maximum distance of the C and N atoms from the mean plane of the ring is in fact 0.219 Å for Fe(OEC)Py and 0.077 Å for Fe(OEC)Ph.

The Fe–N bond lengths are almost the same as those observed in the Fe^{4+} complexes despite the different oxidation state of the metal atom which should

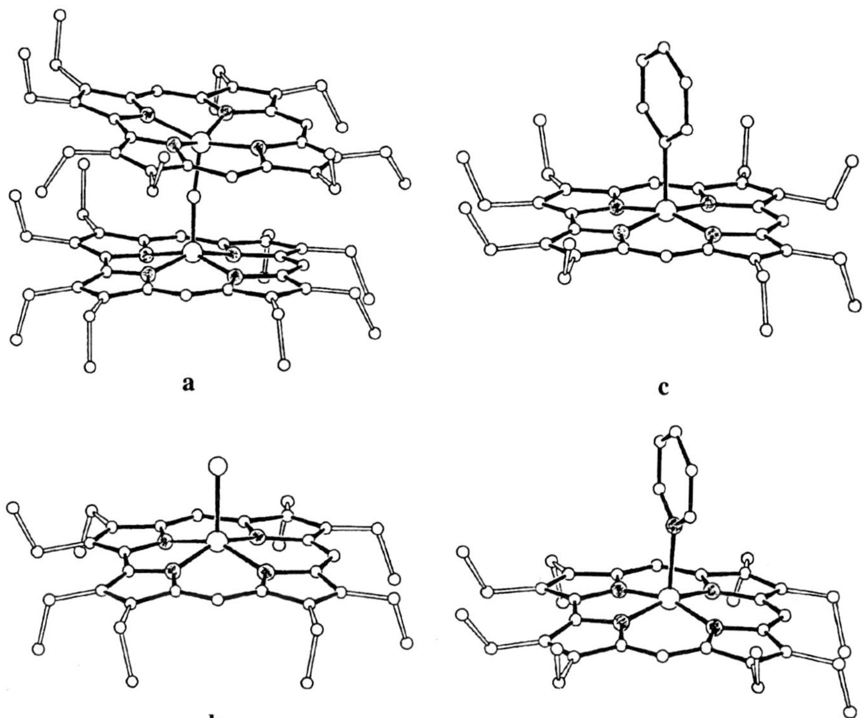

a c b d

Fig. 14a–d. The crystal structures of **a**) $[Fe(OEC)]_2O$, **b**) Fe(OEC)Cl, **c**) Fe(OEC)Ph, **d**) Fe(OEC)Py. Taken from Ref. [27] with permission

Table 4. Selected bond lengths and distances (Δ) for iron corrolates. According to Ref. [27]

Complex	Fe-N[a]	Δ(N4)[b]	Δ(core)[c]
$[Fe(OEC)]_2O$	1.904 Å	0.403 Å	0.427 Å
Fe(OEC)Cl	1.906 Å	0.422 Å	0.533 Å
Fe(OEC)Ph	1.871 Å	0.272 Å	0.318 Å
Fe(OEC)Py	1.893 Å	0.273 Å	0.4425 Å

[a] Average values.
[b] Distance of the iron atom from the mean plane of the four pyrrole nitrogen atoms.
[c] Distance of the iron atom from the mean plane of the ring framework

have produced an elongation towards the distances found in Fe^{3+} porphyrinates. The distance between the metal atom and the pyridine nitrogen atom (2.188 Å) is similar to that reported for iron porphyrinates. Fe(OEC)Py exists in the crystal as a π-π dimer (intermolecular distance 3.6–3.7 Å).

Selected data on the structures of iron derivatives of octaethylcorrole are reported in Table 4.

4.2 Infrared Spectra

The data reported in the literature relative to the IR spectra of metallocorrolates have been used mainly to characterized their axial coordination. Selected data are reported in Table 5.

For instance, the absorptions at 948 cm^{-1} and 964 cm^{-1} have been considered diagnostic of the metal oxygen double bond in the IR spectra of Mo^{5+} and Cr^{5+} derivatives of 2,3,17,18-tetramethyl-7,8,12,13-tetraethyl-corrole H_3(TMTEC) [33] (Fig. 15).

The stretching absorption of the coordinated isocyanides in the spectra of $Rh(OMC)(CNR)_2$ complexes fall around 2200 cm^{-1}. Their energies are lower than those of the corresponding bands in the spectra of analogous bis-isocyanide Rh^{3+} porphyrinates [21]. The explanation that has been proposed is that the higher electronic density of the rhodium atom induced by the corrole ring with respect to the porphyrin enhances the π back donation from the filled *d* orbitals of the metal to the anti-bonding orbitals of the ligand.

The low values of the ν(N≡C) suggest that the coordinated isocyanide cannot undergo nucleophilic attack. Indeed, the possibility of modifying the *p*-tolylisocyanide ligand in a Co^{3+} corrole derivative has been investigated in the past [34]: this compound was stable in refluxing methanol and no modifications or dissociation of the axial ligand were observed.

A similar effect has been observed in the spectrum of $RhCO(OMC)PPh_3$: the carbonyl stretching band is centered at 1985 cm^{-1}, 100 cm^{-1} lower than the corresponding band in the spectrum of RhCO(TPP)Cl [21].

In the case of tin corrolates the IR spectra have been utilized to determine the structure of complexes with different counter ions [25]. The acetate ion behaves as bidentate ligand as confirmed by the low value of the CO stretching vibration and the complex has been suggested to have the structure reported in

Table 5. IR data for metal corrolates

Complex	Vibration	Wave number (cm^{-1})	Reference
CrO(TMTEC)	ν(Cr = O)	964	33
MoO(TMTEC)	ν(Mo = O)	948	33
$RhCO(OMC)PPh_3$	ν(CO ≡ O)	1985	21
Rh(OMC)(*t*-$BuNC)_2$	ν(N ≡ C)	2200	21
$Rh(OMC)(BzNC)_2$	ν(N ≡ C)	2215	21
Sn(OMC)AcO	ν(C = O)	1560	25
Ge(OMC)AcO	ν(C = O)	1572	25
Sn(OMC)Cl	ν(Sn–Cl)	268	25
Ge(OMC)Cl	ν(Ge–Cl)	284	25

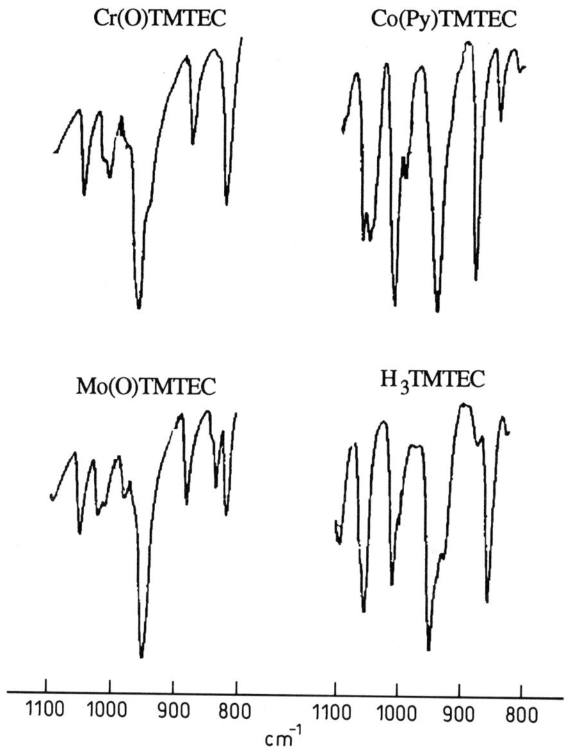

Fig. 15. IR spectra of metallocorrolates. $(TMTEC)^{3-}$ represents the trianion of 2,3,17,18-tetramethyl-7,8,12,13-tetraethylcorrole. Taken from Ref. [33] with permission

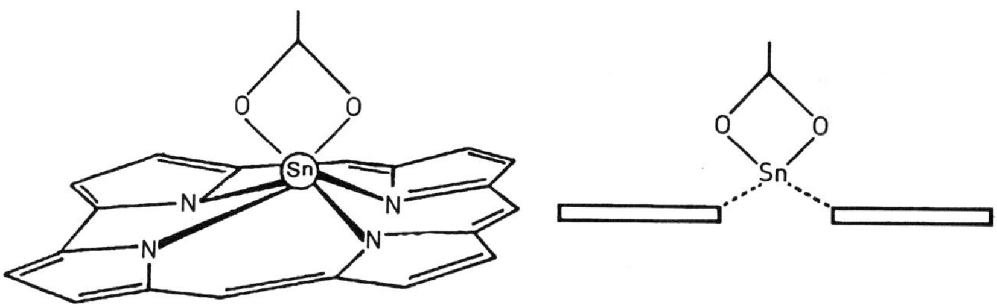

Fig. 16. Coordination geometry of Sn(OMC)OAc. Taken from Ref. [25] with permission

Fig. 16. The tin atom is "pulled" out of the macrocycle plane: such geometry introduces a distortion in the macrocycle that can now be planar or ruffled and the disruption of the classical N–S symmetry of corrole is also reflected by the NMR spectrum of the complex (see Sect. 4.5).

The affinity of Sn corrolate towards different counterions is different from that of Sn porphyrinate: with this latter ligand in fact the favored product is $Sn(P)Cl_2$ and acetate behaves as a monodentate ligand leading to the formation of $Sn(P)(OAc)_2$ complexes [35]. This has been explained by considering the different sizes of the chelating cavities for the two macrocycles and the different binding pattern of the acetate ion in the two cases. Tin derivatives of porphyrins are reported to be those where the major radial expansion of the macrocycle core occurs. The acetate, behaving in the corrole complex as a bidentate ligand and pulling the tin atom out of the corrole plane, decreases the steric constraint due to coordination. The smaller size of the coordination "hole" of corrole has thus been assumed to be the driving force for the preference of acetate.

4.3 Electronic Spectra

The electronic spectra of corroles and metallocorrolates are mainly determined by transitions within the π system of the macrocycle.

Calculations of such spectra have been reported assuming the molecule to be planar with C_{2v} symmetry [36]. The lower symmetry of corrole removes the degeneracy present in porphyrins so that degenerate transitions are no longer expected and more transitions are allowed in the case of corrole than in that of porphyrin. The $7a_2$ and $8b_1$ levels in corrole correspond to the two components of the lowest vacant MO in porphyrins, the $4e_g$ pair (LUMO). Similarly the $6a_2$ and $7b_1$ orbitals correspond to the highest filled porphyrin orbital, a_{1u} and a_{2u} (HOMO). In the case of porphyrins the Q and Soret (thus called because it was discovered in 1883 by Soret in the optical spectrum of hemoglobin) bands arise mainly from the a_{1u}, a_{2u} and e_g pair and the spectra of corroles closely resemble those of porphyrins (Table 6).

Transitions are thus predicted and observed around 570–590 nm as the 0–0 vibronic component of the lowest $\pi \rightarrow \pi^*$ transition and around 530–560 nm as the 0-1 vibronic component of the same transition. The 400 nm absorption is considered to arise from a combination of the second $\pi \rightarrow \pi^*$ transition and its vibrational components.

Table 6. Calculated π molecular orbital (MO) energies (eV) and symmetries for the fully conjugated corrole and porphyrin rings. Both macrocycles have 26 π-electrons. Their net charges are − 3 and − 2, respectively. According to Ref. [36]

Corrole			Porphyrin		
MO No.	Symmetry	Energy	MO No.	Symmetry	Energy
16	$9b_1$	9.320	17	$2a_{1u}$	6.310
15	$8b_1$	6.375	16	$2b_{1u}$	3.779
14	$7a_2$	5.919	14, 15	$4e_g$	2.712
13	$7b_1$	2.347	13	$3a_{2u}$	− 1.362
12	$6a_2$	1.346	12	$1a_{1u}$	− 2.139
11	$6b_1$	− 0.315	10, 11	$3 e_g$	− 3.967
10	$5a_2$	− 0.408	9	$2b_{2u}$	− 4.195

The presence of a Soret band in the optical spectra of all metallocorrolates can then be considered as a proof of their aromaticity.

Data relative to several complexes are reported in Table 7.

The spectra of first-row transition metal complexes are very similar to those of the corresponding metalloporphyrinates. They have been classified as d-type hyper spectra, with additional bands, apart from the Q and Soret ones, in the visible region [24]. As in the case of porpyrinates these bands have been attributed to charge transfer transitions and are reported to be sensitive to solvent and axial coordination.

In the case of the iron complex Fe(OMC) the intensity of the absorption has been related to the existence of stacking phenomena among the macrocyclic units, which are eliminated by interactions with chloride ions or axial ligands [28].

The electronic spectrum of Fe(OMC) in chloroform has been recorded in the presence of several anions. No variation has been observed upon addition of non-coordinating anions such as hexafluorophosphate or tetrafluoborate, while the addition of chloride causes a marked increase in the intensity of the Soret band (370 nm).

It is known [37] that the formation of π molecular complexes between macrocycles, generated by interactions of the aromatic systems, is evidenced by loss of intensity and red shift in the Soret region. Indeed the Soret band in the spectrum of Fe(OMC) shifts from 370 nm in chloroform or benzene to 385 nm in a weakly coordinating solvent such as THF to 395 nm in pyridine, where an axially ligated species is formed.

It is easier for intermolecular interactions to occur in the case of M^{3+} corrolates than in analogous complexes with porphyrins, chlorins or phthalo-

Table 7. Electronic spectra of complexes M(OMC)LL′ or M(OMC)LX

M^{n+}	L	L′	X	λ_{max} (nm)	Reference
Rh^{3+}	PPh_2Me	–	–	550, 410[a]	21
Rh^{3+}	PPh_3	–	–	550, 400[a]	21
Rh^{3+}	PPh_3	CO	–	554, 404[a]	21
Rh^{3+}	$AsPh_3$	–	–	562, 532, 400[a]	21
Rh^{3+}	NEt_3	–	–	580, 566, 400[a]	21
Rh^{3+}	$NHMe_2$	–	–	566, 531, 400[a]	21
Rh^{3+}	*t*-BuNC	*t*-BuNC	–	574, 564, 410[a]	21
Rh^{3+}	BzNC	BzNC	–	578, 566, 410[a]	21
Fe^{3+}	–	–	–	546, 385[b]	28
Mn^{3+}	–	–	–	585, 480, 387[b]	24
Sn^{4+}	–	–	AcO^-	574, 537, 406[c]	25
Sn^{4+}	–	–	Cl^-	566, 532, 409[c]	25
Sn^{4+}	–	–	OH^-	578, 543, 410[c]	25
Ge^{4+}	–	–	AcO^-	560, 524, 403[c]	25
Ge^{4+}	–	–	Cl^-	569, 533, 401[c]	25
In^{3+}	–	–	–	577, 539, 409[c]	25
Zn^{2+}	–	–	PyH^{+}[d]	575, 539, 409[c]	25

[a] Diethyl ether. [b] THF. [c] Chloroform. [d] Positive counterion.

cyanines since no repulsion between electrically charged species has to be overcome in order to bring two or more macrocycles close to each other. A chloride ion binds axially to iron corrolate eliminating molecular stacking. The constant of the equilibrium:

$$[\mathrm{Fe(OMC)}]_n + n\,\mathrm{Cl}^- \leftrightarrows n[\mathrm{Fe(OMC)Cl}]^-$$

has been calculated from the proper Hill plot (Fig. 17). Its low value (K = 309) seems to indicate that the equilibrium is shifted towards the aggregated form (i.e. to the left).

In the case of iron derivatives of corrole the data reported in Table 8 seem to indicate that the presence of absorptions in the low energy region of the spectrum is correlated to axial coordination.

Data relative to the electronic spectra of Co(OMC)PPh_3, Co(OMMPC)PPh_3, Co(5,15-OMDPC)PPh_3, Co(5,10-OMDPC)PPh_3, and Co(OMTPC)PPh_3 complexes are reported in Table 9. All compounds exist in diethyl ether as pentacoordinated species but it has been reported [30] that a sixth ligand can be added to the coordination sphere of cobalt if the complexes are dissolved in a strongly binding solvent such as pyridine. The small red shift observed in going from the *meso*-unsubstituted complex to the triphenyl derivative indicates that the corrole moiety is much less sensitive to substitution than the porphyrin core and that the insertion of phenyl groups at the *meso* positions does not cause severe distortions of the macrocycle.

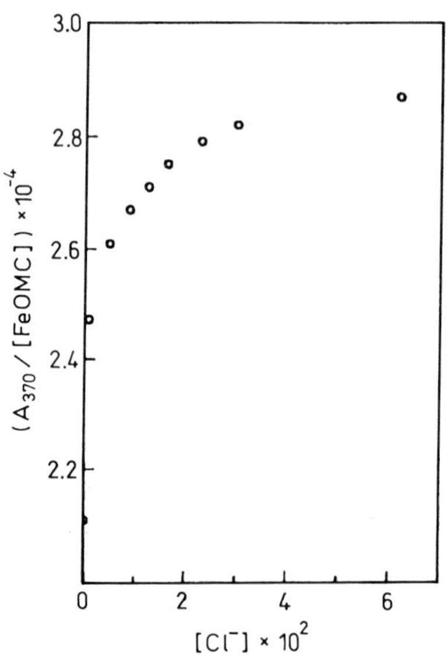

Fig. 17. Effect of chloride concentration on the intensity of the Soret band in the electronic spectrum of Fe(OMC) in chloroform. Taken from Ref. [28] with permission

Table 8. Electronic spectra of iron corrolates. According to Refs. [27, 28].

Complex	λ_{max} (nm)	$\varepsilon \cdot 10^4$	Solvent
$[Fe(OEC)]_2O$	310	5.30	*n*-hexane
	369	10.5	
	537	1.65	
Fe(OEC)Cl	302 sh	2.52	CH_2Cl_2
	338 sh	3.92	
	371	5.74	
	465 sh	1.09	
	516	0.80	
	602	0.30	
Fe(OEC)Ph	251 sh	2.41	CH_2Cl_2
	340 sh	34.3	
	380	6.61	
	507	1.49	
	668	0.23	
Fe(OEC)Py	397	6.72	Pyridine
	482	0.88	
	547	1.52	
	640	0.32	
	670 sh	0.29	
	720	0.20	
Fe(OMC)Py	395	4.04	Pyridine
	546	0.92	
Fe(OMC)	370	2.52	$CHCl_3$

Table 9. Electronic spectra of cobalt corrolates. According to Refs. [30, 31]

Complex	λ_{max} (nm)[a]	$\varepsilon \cdot 10^4$
$Co(OMC)PPh_3$	366	4.47
	531	0.70
	573	1.11
$Co(OMMPC)PPh_3$	367	5.95
	513	1.07
	573	1.22
$Co(5,10\text{-}OMDPC)PPh_3$	375	5.60
	572	0.75
$Co(5,15\text{-}OMDPC)PPh_3$	375	5.60
	572	0.75
$Co(OMTPC)PPh_3$	378	6.03
	574	0.97

[a] Diethyl ether

4.4 Photoelectron Spectra

To study the electronic and geometrical structures of metallocorrolates further, rhodium derivatives have also been investigated by XPS [38]. The study has been extended to rhodium derivatives of *meso*-tetraphenylporphyrin and octaethylporphyrin in order to compare different coordinative structures. Selected data are reported in Table 10.

The spectrum of the free ligand shows two N_{1s} bands which are related to nitrogen hybridization (binding energy 397.7eV N sp^3; binding energy 399.9eV N sp^2) [39]. After complexation there is always a single N_{1s} component shifted to a value which is in between those observed for the free ligand. The general constancy of this energy and the obtained XPS quantitative ratios between metal and nitrogen have been interpreted to be indicative of a coordinative mode of nitrogen to rhodium involving sp^2 N donors.

The rhodium 3*d* binding energies are in the expected range for Rh^{3+} compounds. In general, rhodium corrolates present less positive rhodium atoms than rhodium porphyrinates. The observation is in agreement with the trianionic structure of the corrole moiety that increases electron density on the central metal atom facilitating the extraction of an electron.

The rhodium binding energy is 0.5 eV higher in the case of the bis-isocyanide derivative. This is in favor of the effective π back donation from metal to ligand that is also reflected by the energy of the isocyanide stretching vibration in the IR spectrum of the complex.

XPS data have been also reported in the past [33] for chromium and molybdenum corrolates and discussed with respect to analogous porphyrinates.

The difference between the binding energy of oxochromiun corrolate and porphyrinate increases as shown in the following sequence: Cr $2p_{1/2}$ < Cr $2p_{3/2}$ < Cr $3s_{1/2}$. This has been considered to imply that the energy levels for outer core electrons are affected by valence state more than the levels relative to inner electrons.

4.5 NMR Spectra

Most research involving the NMR of corrins has been performed with vitamin B_{12} and its biologically active intermediates. The purpose of many such studies

Table 10. XPS binding energies (eV) for corrole derivatives. According to Ref. [38]

Compound	N_{1s}	Rh_{3d}
$H_3(OMC)$	397.7 399.9	
$Rh(OMC)PPh_3$	398.9	309.6
$RhCO(OMC)PPh_3$	398.8	309.4
$Rh(OMC)(t\text{-}buNC)_2$	398.4	310.1

has been to understand why the chemical and physical properties of the corrin ring yield the biological functions of vitamin B_{12}. Indeed, it has been proved that one of the fundamental parameters that determine the biological activity of coenzyme B_{12} is the high charge density existing on the cobalt atom and its delocalization over the methine bridges of the corrin ring [40].

Since the interpretation of NMR parameters in complex biological systems is considerably facilitated by the use of model compounds it is evident that a proper understanding of the metal-ligand interactions in the synthetic counterparts of vitamin B_{12}, which are the subject of this review, can yield additional details and help clarify the role of corrinoids in Nature.

4.5.1 Paramagnetic Derivatives

In order to characterize the ligand binding properties in solution of metallocorrolates a detailed NMR study has been carried out on the paramagnetic derivative Fe(OMC) [28]. Evidence has been obtained that indicates a substantial difference in the ligand field effects between corrole and other tetrapyrrolic macrocycles.

NMR provides a powerful tool for the investigation of paramagnetic metal complexes: paramagnetism allows the resolution of magnetically non equivalent environments that cannot be resolved in a comparable diamagnetic system. Furthermore the large expansion of the chemical shift range of the observed resonances (the isotropic shift) permits the characterization of fast dynamic processes.

The analysis of the nature of the isotropic shift which can be due to dipolar (through space) or contact (through bond) interaction also allows the characterization of the metal-macrocycle bonding.

Figure 18 (trace a) shows the spectrum of Fe(OMC) in chloroform. Assignments have been performed by a variety of arguments: chemical shifts, relaxation rates etc.

The most downfield shifted resonances (labeled M_1 and M_2 in Fig. 18) have been attributed to the *meso* protons because of their proximity to the paramagnetic center. The relative intensity of the integral ratio is in agreement with the existence of a C_2 symmetry axis. On the basis of symmetry, four resonances are expected for the eight methyl groups. The observation of eight resonances has not been attributed to the disruption of symmetry but to the existence of different chemical species in solution. Traces b and c in Fig. 18 show the spectra obtained by addition of silver nitrate or excess chloride ions to the original solution. With time, spectrum b tends to go back to the original one (trace a) and the second species is regenerated. This behavior has been ascribed to the oxidative decomposition of chloroform which is known to produce hydrogen chloride or chloride ions:

$$O_2 + 2\,CHCl_3 \xrightarrow{h\nu} 2\,HCl + 2\,COCl_2$$

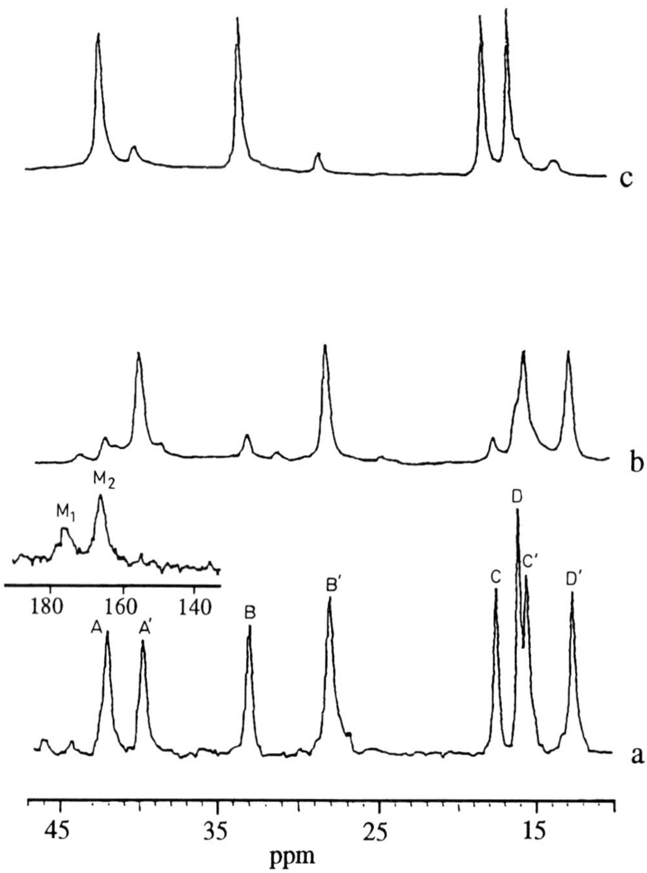

Fig. 18a–c. ^{1}H NMR spectra of Fe(OMC) in chloroform. *Inset* (δ 130–190 ppm) refers to spectrum (*a*). (*a*) in C^2HCl_3; (*b*) resulting from elimination of chloride by treatment of the sample with $AgNO_3$; (*c*) the sample of (*a*) after addition of excess NEt_3BzCl. Taken from Ref. [28] with permission

The two species have thus been related to a binding equilibrium with chloride ions.

A saturation transfer experiment demonstrated that the multiplicity of resonances arises from an equilibrium between different compounds: macrocycles where a chloride ion is bound axially to iron and aggregates where the axial position is essentially free and the electronic π system gives rise to stacking interactions (see Sect. 4.3).

The existence of an axially ligated species has been also demonstrated by preparation in situ of a *t*-butylisocyanide derivative of Fe(OMC): the NMR spectrum of such derivative is reported to be almost identical to that of the chloride bound complex.

The spectral properties of Fe(OMC) in chloroform are summarized in Table 11. Relaxation rates are consistent with a low spin state for iron in this complex.

Table 11. ^1H NMR spectral properties of Fe(OMC) in chloroform. According to Ref. [28]

Label[a]	Intensity[b]	δ (ppm)	T_1 (ms)	T_2 (ms)
M_1	1	175.1	c	0.8
M_2	2	165.6	c	1.1
A	3	41.9	20	12
A′	3	39.7	20	10
B	3	31.4	22	14
B′	3	28.2	28	11
C	3	18.1	28	15
D	3	16.6	37	c
C′	3	16.1	26	c
D′	3	13.3	21	13

[a] See Figure 18.
[b] Intensity as the number of protons.
[c] Not reported

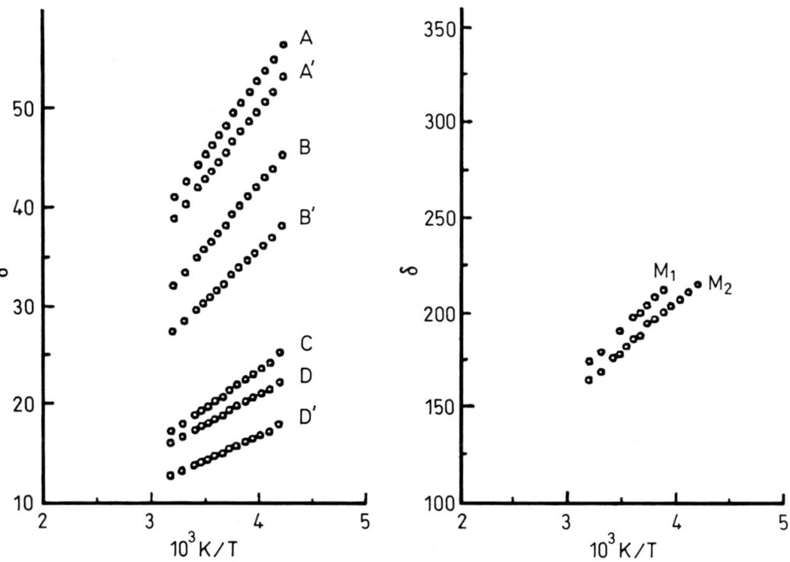

Fig. 19. Temperature dependence of the isotropic shifts of the resonances due to the methyl groups and *meso* protons of Fe(OMC) in chloroform. *Labels* refer to the assignments reported in Table 11. Taken from Ref. [28] with permission

The magnetic moment, determined in the presence of excess chloride ions in order to have a single species in solution, resulted to be 1.7 BM, confirming the low spin, S = 1/2, state for the metal atom.

The linearity of the Curie plots (chemical shifts versus reciprocal temperature) for all the resonances of the NMR spectrum of Fe(OMC) in chloroform demonstrated the existence of a pure spin species where contact mechanisms are the main contribution to the isotropic shift (Fig. 19).

Fe^{3+} derivatives of porphyrins and chlorins are all high spin in the absence of N-donor axial ligands and in chlorinated solvents [41, 42]: the low spin state observed in the case of iron corrolate demonstrates that the ligand field strength of corrole is very different from that of other macrocycles.

The use of pyridine as solvent dramatically alters the spectral properties of the iron complex of octamethylcorrole. Such a spectrum is reported in Fig. 20 and pertinent data in Table 12. The three resonances (A–C) observed at low field have been attributed to the methyl substituents, the fourth resonance probably being located in the diamagnetic region obscured by the solvent resonances.

The resonances at high-field (D, E) have been attributed to the residual protons of deuteriated pyridine axially ligated to iron. This assignment should however be corrected on the basis of the recent data reported by Vogel et al. [27] and the resonances should be attributed to the *meso*-protons of the macrocycle (see Table 13).

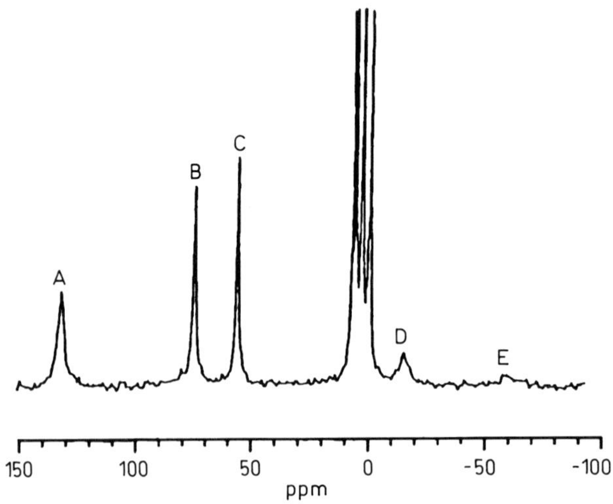

Fig. 20. ^1H NMR spectra of Fe(OMC) in pyridine. *Labels* refer to the assignments discussed in Sect. 4.5.1. Taken from Ref. [28] with permission

Table 12. ^1H NMR spectral properties of Fe(OMC) in pyridine. According to Ref. [28].

Label[a]	Intensity[b]	δ (ppm)	T_1 (ms)	T_2 (ms)
A	6	131.0	1.6	1.3
B	6	75.0	4.8	3.2
C	6	56.4	6.1	3.9

[a] See Fig. 20.
[b] Intensity as the number of protons

Table 13. ^{1}H NMR data (δ, ppm) of iron derivatives of H_3OEC. According to Ref. [27]

	Fe(OEC)Cl[a]	Fe(OEC)Ph[a]	Fe(OEC)Py
5,15-H	177	54.5	– 15.9[b]
10-H	189	49.3	– 62.2[b]
ring-CH_2	– 5.7/29.9	– 8.4/98.1	1.9/67.7[b]
ring-CH_3	– 0.3/ – 2.5	3.0/7.3	3.2/3.8[b]
3,5-Ph		– 4.0	
4-Ph		– 77	
2,6-Ph		– 153.6	
α-Py			114[c]
β-Py			77.1[c]
γ-Py			– 3.3[c]

[a] Deuteriated chloroform solutions.
[b] Deuteriated pyridine solution.
[c] Carbon disulfide solution

The magnetic moment determined for Fe(OMC) in pyridine resulted to be 4.3 BM and such a value together with curvatures present in the Curie plots have been explained by the existence of a spin equilibrium between the $S = 1/2$ and $S = 5/2$ spin states. Considering that the only mixed spin states in iron porphyrinates are observed when the axial positions are occupied by a pyridine and a water molecule or when the iron is displaced from the macrocycle plane, a mixed axial ligation or a pentacoordinated environment have been postulated for iron corrolate in pyridine. However, the large spread of chemical shifts suggests that a strong interaction occurs between the metal ion and the π system of the macrocycle. It is known that the ferric ion in hemin chloride dissolved in pure pyridine is slowly reduced to a bis(pyridine)-Fe^{2+} porphyrin complex [43]. It cannot, therefore, be excluded that similar phenomena occur in the case of iron corrolates with formal charge-transfers involving the metal and the ligand generating delocalization of the electron density over the aromatic system of the corrole ring (see Sect. 3.1). Further studies will be necessary to fully characterize these complexes.

Data relative to the ^{1}H NMR spectra of iron derivatives of H_3OEC are reported in Table 13. The spectra of the mononuclear complexes Fe(OEC)L (L = Cl, Ph, Py) are typical of paramagnetic species with broad unresolved signals distributed over a large spectral width.

The large difference in chemical shift observed for corresponding resonances in the spectra of Fe(OEC)Cl and Fe(OEC)Ph are indicative of a different electronic distribution in the two species. The two complexes have very similar magnetic susceptibilities ($\mu = 2.97$ BM and 2.89 BM, respectively).

The pentacoordinated environment assumed for Fe(OMC)Py is confirmed by the data relative to Fe(OEC)Py: no tendency to form a bis-pyridine adduct has been observed for this latter complex as indicated by the substantial invariance of the spectra recorded in pyridine or carbon disulfide.

4.5.2 Diamagnetic Derivatives

All resonances in the proton NMR spectra of diamagnetic metallocorrolates show a strong upfield shift due to anisotropic effects caused by the macrocycle ring current [44]. This is another demonstration of the aromatic character of the corrole ring [21]. Spectral properties of several derivatives of octamethylcorrole are reported in Table 14.

No differences have been observed in the pattern of resonances due to the peripheral methyl substituents in going from penta- to hexacoordinated species. The resonance at lower field can be probably attributed to the 2,18-CH_3 since electron density is known to be higher in these positions [45].

Table 14. ^{1}H NMR data of some diamagnetic metallocorrolates. According to Ref. [25].

Complex	δ (ppm)[a]	Intensity[c]	Assignment
Sn(OMC)AcO	10.56	3	*meso*-H
	3.74	24	ring CH_3
	– 1.36	3	AcO^- CH_3
Sn(OMC)AcO	10.22[b]	2	5,15-H
	10.00[b]	1	10-H
	3.80[b]	6	2,18-CH_3
	3.75[b]	18	ring CH_3
	– 1.40[b]	3	AcO^- CH_3
Sn(OMC)Cl	10.10	2	5,15-H
	9.92	1	10-H
	3.95	6	2,18-CH_3
	3.75	18	ring CH_3
Ge(OMC)AcO	9.75	3	*meso*-H
	3.78	6	2,18-CH_3
	3.63	6	8,12-CH_3
	3.58	12	3,7,13,17-CH_3
	– 1.35	3	AcO^- CH_3
Ge(OMC)Cl	9.74	2	5,15-H
	9.66	1	10-H
	3.83	6	2,18-CH_3
	3.61	18	ring CH_3
In(OMC)	10.32	3	*meso*-H
	3.70	24	ring CH_3
$[Zn(OMC)]^-[PyH]^+$	10.04	2	5,15-H
	9.98	1	10-H
	7.12–6.68	6	PyH^+
	3.63	18	ring CH_3
	3.52	6	ring CH_3

[a] 80 MHz spectra. All resonances appear as singlets. No linewidth data have been reported.
[b] 400 MHz spectrum.
[c] Intenisty as the number of protons

Table 15 reports NMR data for rhodium derivatives of octamethylcorrole [21].

The resonances due to the *meso* protons appear as two signals in the spectra of all rhodium complexes. In most cases their relative intensity is 2:1 (5,15-H and 10-H) with the 5,15-H resonating at lower field. However, when the axial ligand is PPh_3 or $AsPh_3$ the situation is reversed and it is the 10-H which resonates at lower field.

Since no variations of the spectra have been observed between 300 and 340 K, in a concentration range between 10^{-2} and 10^{-4} M, the existence of conformational equilibria has been ruled out. The variation of the intensities of the *meso*-proton resonances has then been attributed to a contribution of the shielding effect of the phenyl rings of the axial ligands to that of the macrocycle.

The *ortho*-protons of the phenyl rings of the triphenylphosphine ligand in the spectra of $Rh(OMC)PPh_3$ and $Rh(OMC)PPh_3(CO)$ resonate at 4.58 and 4.50 ppm respectively. Such a small difference can be explained if one assumes that in the hexacoordinated environment the metal atom will be closer to the macrocycle plane the ring current of which would then be felt more strongly by the protons of the axial phosphine ligand. Similarly, the resonances due to the methyl substituents of the corrole ring show the highest values of chemical shifts in the spectra of the symmetric bis-isocyanide derivatives, in which the metal atom is expected to be located in the macrocycle plane.

Table 15. ^{1}H NMR data of Rh(OMC)LL′ complexes. According to Ref. [21]

L	L′	5,15-H	10-H	Ring Me[a]	L
PPh_2Me	–	9.49	9.45	3.53 (6H) 3.49 (6H) 3.39 (12H)	6.87–6.20 (6H) 4.39–4.15 (4H) 1.94 (6H)
PPh_3	–	9.39	9.41	3.45 (6H) 3.39 (6H) 3.33 (12H)	6.62–6.38 (9H) 4.71–4.45 (6H)
PPh_3	CO	9.09	9.14	3.42 (6H) 3.35 (6H) 3.27 (12H)	6.96–6.52 (9H) 4.65–4.35 (6H)
$AsPh_3$	–	9.40	9.48	3.41 (12H) 3.33 (12H)	6.89–6.35 (9H) 4.81–4.7 (6H)
NEt_3	–	9.22	9.06	3.65 (6H) 3.52 (6H) 3.48 (12H)	– 2.12 (9H) – 3.54 (6H)
$NHMe_2$	–	9.23	9.04	3.67 (6H) 3.55 (6H) 3.49 (12H)	– 3.37 (6H) – 5.80 (1H)
t-buNC	*t*-buNC	9.96	9.74	4.04 (6H) 3.91 (6H) 3.81 (12H)	– 0.96 (18H)
BzNC	BzNC	10.00	9.79	4.12 (6H) 3.96 (6H) 3.87 (12H)	7.04–6.83 (6H) 4.62–4.45 (4H) 2.08 (4H)

[a] All resonances appear as singlets. No linewidth data have been reported

Another example of how spectral properties have been related to the geometry of the complexes is the case of tin and germanium corrolates [25].

In the 80 MHz ^{1}H-NMR spectrum of Sn(OMC)Cl, and in the analogous germanium complex, the *meso*-protons appear as two signals while in the spectra of the acetate derivatives a single resonance has been observed. No linewidth data have been reported and the results have been explained considering that the behavior of acetate as a bidentate ligand, discussed in Sect. 4.2, would introduce distortions into the molecule causing accidental isochrony of the *meso*-proton resonances.

In the 400 MHz ^{1}H-NMR spectrum of Sn(OMC)OAc, however, the resonances due to the 10-H and 5,15-H are resolved as well as those due to ring-methyls (see Table 14) demonstrating that the pattern observed at 80 MHz is indeed generated by the accidental superimposition of the signals.

No difference has been observed in the 80 MHz or 400 MHz ^{1}H-NMR spectra of Sn(OMC)Cl.

The effect of peripheral substituents in the ^{1}H-NMR spectra of diamagnetic corrolates has been discussed for Co(OMC)PPh_3 and the series of *meso* mono-, di- and triphenyl cobalt corrolates [30, 31].

In the case of porphyrins different substitution patterns induce relevant electronic variations: the energy level of the two HOMOs of the porphyrin π system (a_{1u} and a_{2u}) are in fact reversed in going from a *meso*-substituted porphyrin, such as tetraphenylporphyrin, to a β-substituted one, such as octaethylporphyrin [44]. Electronic variations may also derive from conformational changes: Smith and coworkers have reported hybrid compounds substituted both at *meso* and β positions, the Zn^{2+} complexes of 2,3,7,8,12,13,17,18-octaethyl- and 2,3,7,8,12,13,17,18-octamethyl-5,10,15,20-tetraphenylporphyrin [46] (see Sect. 3.3). These complexes have a nonplanar conformation caused by the steric interactions of the substituents. Theoretical calculations indicate that such conformational changes cause a differential shift in the energies of the HOMO and LUMO and so modulate the electronic structure of the macrocycles.

Data relative to the 400 MHz ^{1}H-NMR spectra of cobalt corrolates are reported in Table 16. The strong shift due to the macrocycle ring current demonstrates that the presence of the *meso*-phenyl substituents does not cause the loss of aromaticity. The *meso*-protons of the monophenyl derivative resonate at the same chemical shift value observed for the *meso*-unsubstituted complex.

The resonances due to the protons of the axial triphenylphosphine ligand also demonstrate the shielding effect of the aromatic macrocycle. Such effect decreases when going from Co(OMC)PPh_3 to Co(OMTPC)PPh_3 indicating that in the latter compound the axial ligand is at a larger distance from the macrocycle plane probably because of steric constraints introduced by the presence of the three phenyl substituents.

The different patterns shown by the resonances due to the peripheral methyl groups have been considered to depend on the number of *meso*-phenyl substituents and in the case of the diphenyl derivatives have been used to identify the

Table 16. 400 MHz ^{1}H NMR data of cobalt corrolates. According to Refs. [30, 31]

Complex	δ (ppm)	Intensity[a]	Assignment
Co(OMC)PPh_3[b]	9.42, *s*	1	10-*meso*-H
	9.03, *s*	2	15-*meso*-H
	7.00, *t*	3	*p*-PPh_3
	6.66, *t*	6	*m*-PPh_3
	4.70, *t*	6	*o*-PPh_3
	3.23, *s*	18	3,7,8,12,13,17-CH_3
	3.15, *s*	6	2,18-CH_3
Co(OMMPC)PPh_3[b]	9.03, *s*	2	5,15-*meso*-H
	7.85, *d*	1	*o*-10-Ph
	7.62, *t*	1	*m*-10-Ph
	7.57, *t*	1	*p*-10-Ph
	7.45, *t*	1	*m'*-10-Ph[c]
	7.13, *t*	1	*o'*-10-Ph[c]
	7.03, *t*	3	*p*-PPh_3
	6.69, *t*	6	*m*-PPh_3
	4.79, *t*	6	*o*-PPh_3
	3.25, *s*	6	3,17(or 7,13)-CH_3
	3.15, *s*	6	7,13(or 3,17)-CH_3
	3.04, *s*	6	2,18-CH_3
	2.15, *s*	6	8,12-CH_3
Co(5,10-OMDPC)PPh_3[c]	9.40, *s*	1	15-*meso*-H
	7.86–7.20, *m*	10	5,10-*meso*-Ph
	6.83, *t*	3	*p*-PPh_3
	6.63, *t*	6	*m*-PPh_3
	5.26, *t*	6	*o*-PPh_3
	3.25, *s*	3	18-CH_3
	3.20, *s*	3	2-CH_3
	3.14, *s*	3	17-CH_3
	3.11, *s*	3	13-CH_3
	2.43, *s*	3	3-CH_3
	2.40, *s*	3	12-CH_3
	2.39, *s*	6	7,8-CH_3
Co(5,15-OMDPC)PPh_3[c]	9.80, *s*	1	10-*meso*-H
	7.87–7.42, *m*	10	5,15-*meso*-Ph
	6.78, *t*	3	*p*-PPh_3
	6.56, *t*	6	*m*-PPh_3
	5.18, *t*	6	*o*-PPh_3
	3.15, *s*	12	2,8,12,18-CH_3
	2.48, *s*	12	3,7,13,17-CH_3
Co(OMTPC)PPh_3[c]	7.91–7.43, *m*	15	*meso*-Ph
	6.83, *t*	3	*p*-PPh_3
	6.64, *t*	6	*m*-PPh_3
	5.36, *t*	6	*o*-PPh_3
	3.14, *s*	6	2,18-CH_3
	2.37, *s*	12	3,7,13,17-CH_3
	2.33, *s*	6	8,12-CH_3

[a] Intensity as the number of protons.
[b] Deuteriated chloroform solutions.
[c] The prime indicates the *meso*-phenyl protons located on the same side of the axial ligand with respect to the macrocycle plane.
[d] Deuteriated benzene solutions.

two isomers. In the spectrum of $Co(OMMPC)PPh_3$ the eight peripheral methyl groups give rise to four signals. Symmetry considerations and the shielding effect of the 10-phenyl group made it possible to assign the resonance at 2.15 ppm to the 8,12-CH_3 and the one at 3.04 ppm to the 2,18-CH_3. In the spectrum of $Co(OMTPC)PPh_3$ the methyl groups appear as three signals at $\delta = 3.03$, 2.05, 2.00 ppm of relative intensity 6:12:6; again on the basis of symmetry considerations such signals have been assigned to the 2,18-CH_3, 3,7,17,18-CH_3 and to the 8,12-CH_3 substituents, respectively. In the case of $Co(OMC)PPh_3$ two signals are present in the high field region of the spectrum ($\delta = 3.23$, 3.15 ppm). Their relative intensity (18:6) and the comparison with the spectra of the *meso*-substituted complexes allowed to assign the signal at 3.15 ppm to the 2,18-methyl groups.

A general feature of *meso*-phenyl substituted Co^{3+} corrolates is then that the planarity of the macrocyclic ligand is maintained in solution: the pattern shown by the resonances due to the peripheral methyl groups in fact are indicative of the existence of a C_2 symmetry axis the direct pyrrole–pyrrole bond typical of a planar corrole skeleton, confirmed also by the signals due to the *meso*-protons, if present.

This is another result that reveals the peculiarity of the corrole ligand with respect to other macrocycles: similarly substituted porphyrins have very different properties. Despite the presence of a direct pyrrole–pyrrole link then, corrole seems to be less sensitive to steric constraints than porphyrin. The crystal structure of $Co(OMTPC)PPh_3$ (see Sect. 4.1) confirms this analysis.

In the ^1H NMR spectrum of $[Fe(OEC)]_2O$ the presence of well defined signals with small linewidth indicated that the complex is diamagnetic [27] because of strong antiferromagnetic coupling between the two iron atoms. The magnetic susceptibility measurement ($\mu = 0.69$ BM) of the complex in the solid state confirmed this analysis. The observed high field shift of the resonances (no less than 2.7 ppm) cannot be explained on the basis of magnetic anisotropy of the ligands (as in the case of metalloporphyrin dimers) and demands further investigations [27].

4.6 Electrochemistry

It is well known that the oxidation state of the cobalt atom changes in the enzymatic B_{12} dependent reactions. The redox chemistry of cobalt corrinoids received a great deal of attention in the seventies. The Co^{3+}/Co^{2+} reaction of different corroles was investigated in the past although no mechanistic details have been reported. The resulting chemistry has already been reviewed [11].

Since then, the metal derivatives that have been studied include Mo and Cr complexes which have been electrochemically investigated by Murakami et al. [47, 48] during their studies on one-dimensional electric conductors and, more recently, a detailed study of the electrochemistry of Co^{3+} and Rh^{3+} has been reported [49].

Cyclic voltammetry and controlled-potential electrolysis are the techniques that have been used to investigate the electrochemistry of oxo-chromium and oxo-molybdenum corrolates. The data have been related to those obtained for similar porphyrin complexes. Redox potentials are reported in Table 17.

For both complexes, reversible metal-centered one-electron oxidations and reductions have been observed. The products of such redox processes have been examined by monitoring the EPR and electronic spectra obtained by controlled-potential electrolysis and, in the case of the molybdenum complex, have been identified as $[Mo(IV)O(TMTEC)]^+$ and $[Mo(VI)O(TMTEC)]^-$ ions.

No ligand centered reduction has been observed in a cathodic region up to − 2.0 V (vs SCE): the HOMO–LUMO energy separation in corrole resulted then to be larger than 3.3 V, i.e. much greater than that present in porphyrins. This is consistent with the lower skeletal symmetry of the corrole structure with respect to porphyrin which is expected to increase the separation between HOMO and LUMO [46].

The oxochromium complex shows a very similar electrochemistry: it also undergoes two metal centered reversible one-electron redox processes and one reversible oxidation of the ligand.

The large differences in metal-centered redox potentials between chromium and molybdenum have been attributed to the existence of a greater interaction between the Cr d_{xy} orbital and the σ and/or π orbitals of the corrole ligand. Such interaction would in fact reduce d electron density at the metal. Another factor considered to cause the potential difference has been a strong covalent interaction between chromium and N-donor ligands.

The electrochemical data for $Rh(OMC)PPh_3$ and $Co(OMC))PPh_3$ in several organic solvents are reported in Table 18.

Both complexes undergo up to three one-electron oxidations. The first two processes are reversible at all scan rates while the third one is irreversible at low scan rates.

The oxidized complexes have been spectroscopically characterized. Their formation is straightforward and it has been accounted for by the following

Table 17. Half wave potentials (in V vs SCE) of oxochromium(V) and oxomolybdenum(V) macrocyclic complexes in dichloromethane. According to Refs. [47, 48]

Complex	Metal		Ligand	
	$E_{1/2}$ (ox)	$E_{1/2}$ (red)	$E_{1/2}$ (ox)	$E_{1/2}$ (red)
MoO(TMTEC)	0.70	− 0.72	1.30	< − 2.00
CrO(TMTEC)	0.63	− 0.33	1.30	< − 2.00
MoO(TPP)OMe	> 1.70	− 0.74	1.50	− 1.14
MoO(TPP)Cl	1.30	− 0.06	1.50	− 1.11

Table 18. Half wave potentials/V vs SCE of $M(OMC)PPh_3$. According to Ref. [49]

M	Solvent	Oxidation			Reduction	
		1st	2nd	3rd	1st	2nd
Rh^{3+}	CH_2Cl_2	0.15	0.61	1.43	− 1.39	− 1.39
	PhCN	0.21	0.66	1.42	− 1.27	− 1.34
	THF	0.23	0.67		− 1.20	− 1.29
	DMF	0.16	0.61		− 1.21	− 1.31
Co^{3+}	CH_2Cl_2	0.18	0.80	1.68	− 0.85	− 1.90
	PhCN	0.19	0.76	1.54	− 0.86	− 1.92
	THF	0.30	~ 0.80[a]		− 0.72	− 1.84[b]
	DMF	0.25	0.83		− 0.79	− 1.83[b]

[a] Uncertainty in the value due to the presence of two overlapping peaks.
[b] Value obtained at T = 203 K

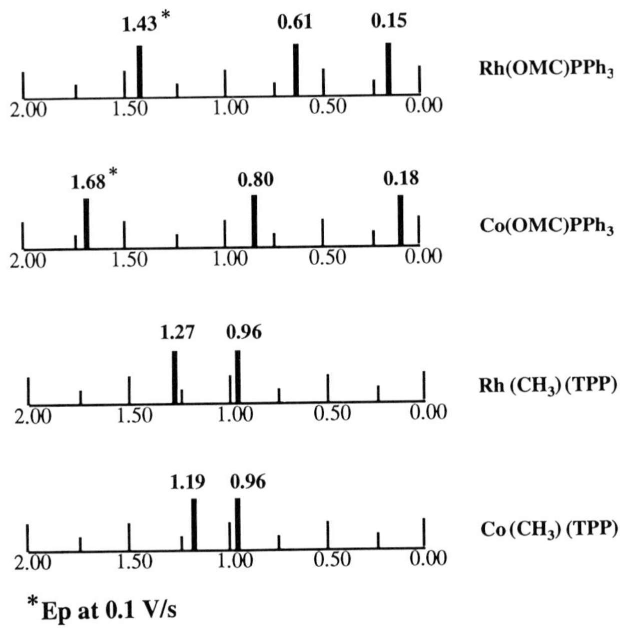

Fig. 21. Half wave potentials of the oxidations of $Rh(OMC)PPh_3$, $Co(OMC)PPh_3$, $Rh(TPP)CH_3$ and $Co(TPP)CH_3$ in dichloromethane. Taken from Ref. [49] with permission

reactions where all the electrons are abstracted from the corrole π ring system:

$$M^{III}(OMC)PPh_3 \rightleftarrows [M^{III}(OMC)PPh_3]^+ + e^-$$

$$[M^{III}(OMC)PPh_3]^+ \rightleftarrows [M^{III}(OMC)PPh_3]^{2+} + e^-$$

$$[M^{III}(OMC)PPh_3]^{2+} \rightleftarrows [M^{III}(OMC)PPh_3]^{3+} + e^-$$

The first oxidation occurs at potentials more negative than those reported for the oxidations of similar metalloporphyrinates as shown graphically in

Fig. 21. This trend is consistent with the XPS data previously discussed (see Sect. 4.4) and shows the higher basicity of the corrole ring which has been attributed to a simple inductive effect.

On the reduction side the behavior of the two complexes becomes different: cobalt shows two one-electron reductions, the first one being irreversible and the second reversible at all scan rates, while rhodium exhibits an irreversible two-electron reduction associated with two reoxidation peaks.

On the basis of spectroelectrochemistry it has been proposed that cobalt corrolate is reduced with a mechanism similar to that occurring in the case of cobalt porphyrinates as shown in Fig. 22 where the cyclic voltammograms in the four non-aqueous solvents investigated are also shown.

The irreversible reaction has been ascribed to a variation of geometry which would occur upon dissociation of the axial ligand. The authors of Ref. [50] reported the generation of a four coordinated Co^{2+} complex with D_{4h} symmetry demonstrated by EPR.

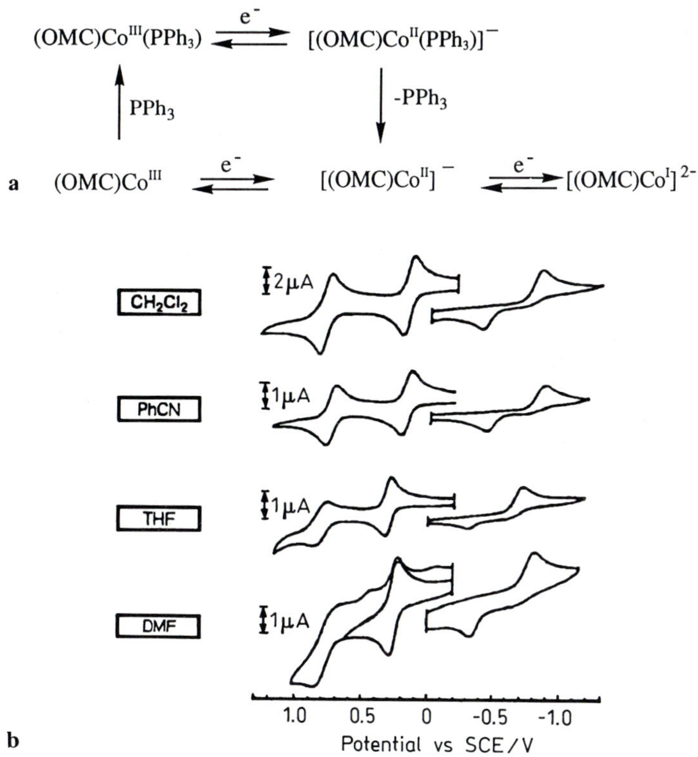

Fig. 22. a) The mechanism of the electroreduction of $Co(OMC)PPh_3$. **b**) Cyclic voltammograms of $Co(OMC)PPh_3$ in four nonaqueous solvents at room temperature. Taken from Ref. [49] with permission

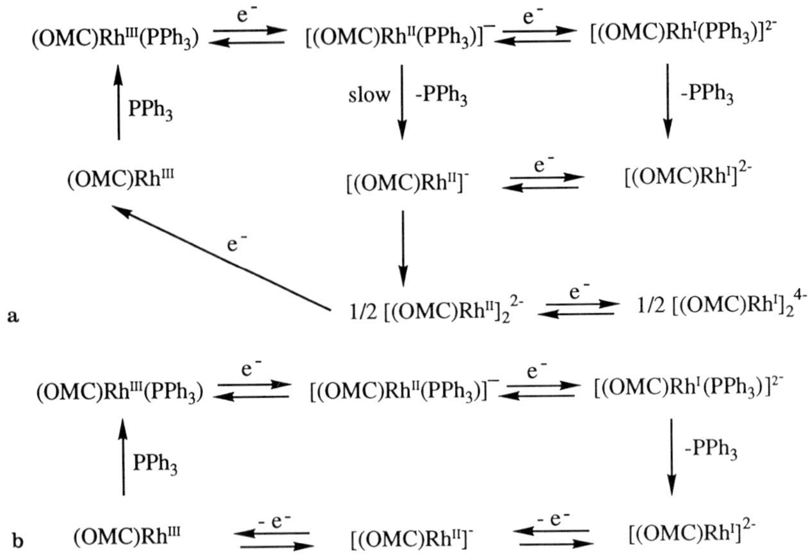

Fig. 23a, b. The mechanism of the electroreduction of Rh(OMC)PPh_3. a) all conditions except those in (b). b) high phosphine concentration, fast scan rate. Taken from Ref. [49] with permission

The reduction peaks do not vary if the electrochemistry is carried out in the presence of excess PPh_3. It is known that Co^{2+} porphyrinates can coordinate donor molecules along the z axis and the lack of occurrence of such reaction in the case of corroles has been attributed to the negative charge of the electrogenerated Co^{2+} complex.

The electrochemical and spectroelectrochemical data for rhodium corrolate have been explained with the existence of monomers and dimers in solution and the reductions have been proposed to occur according to the mechanism shown in Fig. 23a.

When the reduction has been carried out at high concentration of PPh_3, dimerization is prevented and the mechanism proposed is that reported in Fig. 23b.

5 Syntheses of Corrinoids

In the laboratory, as well as in Nature, corrinoid macrorings must be constructed from pyrroles. The complicated substitution patterns and the various unsaturation levels that can be introduced in the corrinoid structures represented a serious problem for the organic synthetic chemists and have stimulated many investigations in the past.

This section will not be concerned with the detailed description of the synthetic methods leading to the appropriate precursors: we will limit our attention to the crucial step of the synthesis of corrinoids, i.e. the formation of the tetrapyrrolic ring. The corrinoid macrocycle has been synthesized following two different procedures: the first one involves the cyclization of a proper linear precursor, while the second involves ring contraction of a porphyrinoid structure.

5.1 Cyclization of Linear Precursors

This method affords the macrocycle by final formation of the direct link between pyrroles A and D from an appropriate linear tetrapyrrole. The proper reduced bilin must be synthesized in order to have a specific substitution pattern of the corrinoid structure.

The cyclization reaction can be carried out in different conditions depending on the oxidation level of the desired corrinoid. Now all members of the family of such tetrapyrroles are available. Their structures have been reported in Fig. 7 and a list of the abbreviations used for the different ligands is reported in Sect. 2.3.

The synthetic methods leading to corrinoids with different degrees of unsaturation will be described separately.

5.1.1 Synthesis of Octadehydrocorrins

The presence of a metal salt is generally necessary to achieve the macrocycle: some corrinoids have been isolated only as metal complexes since the metal-free ligand is not stable enough to be isolated. The reasons for such instability have not been clarified, but it is noted to increase with the unsaturation level of the macrocycle.

The first example reported in the literature is the cyclization of dihydrobilin to octadehydrocorrin [51–54]. The reaction is catalyzed by the presence of nickel or cobalt salts. As in the case of corrole and its metal complexes such ring closure reaction has been carried out in alcoholic solution, it is oxidative and base catalyzed. It has been demonstrated that the formation of the corrin ring is part of an equilibrium where the oxidative ring closure is coupled with a reductive ring opening reaction [55].

The 1,19-substituted octadehydrocorrin complexes have been isolated as crystalline materials and characterized. In the case of the 1-substituted species only the nickel complex turned out to be stable, while the cobalt derivative showed a very high reactivity and all attempts of purification afforded a mixture of compounds [56, 57].

The same complexes have also been prepared by cyclization of tetrahydrobilins in similar experimental conditions [3, 58, 59].

The chemistry of Ni and Co complexes of octadehydrocorrin has been reviewed in the past [3, 8, 10], further developments will be discussed in Sect. 6.

(1,19-dicarboxyethyl-2,3,7,8,12,13,17,18-octamethyl-octadehydrocorrinato)-palladium perchlorate [60] and (1,19-dicarboxyethyl-2,3,7,8,12,13,17,18-octamethyl-octadehydrocorrinato)platinum bromide [61] have been synthesized by cyclization of dihydrobilins and represent the only examples of complexes with metals other than Ni or Co. Attempts have been made to prepare complexes with different metals but they have been unsuccessful.

When copper salts are used to catalyze the cyclization of dihydrobilin the reaction product is the corresponding Cu(II) porphyrin which can be demetalated if dissolved in strong acidic media: this method has been used for the synthesis of porphyrins with asymmetrical substitution patterns [62].

Other metal salts, such as Cr^{3+}, Rh^{3+} or Ru^{3+} derivatives, catalyze the cyclization of dihydrobilin to porphyrin [63]: differently from copper, these metals afford the metal-free ligand and this seems to be a suitable route for the preparation of porphyrins with acid-labile substituents, which in the previous method suffer during the removal of the copper atom. On the other hand, dihydrobilin also cyclizes to porphyrin when heated in 1,2-dichlorobenzene or when treated with bases although in lower yields [3, 8, 10]. The presence of nickel or cobalt salts seems to be necessary to drive the cyclization reaction towards the formation of the octadehydrocorrin structure.

Nickel or cobalt derivatives of 1,19-dimethyl-octadehydrocorrinates may be hydrogenated and the corresponding complexes of reduced corrins are obtained, the degree of unsaturation introduced in the macrocycle depending on both the reaction procedures and β substitution patterns [51–54]. Thus when 1,19-dimethyl-octaalkyl-octadehydrocorrinate is vigorously hydrogenated over Ni-Raney, the corresponding didehydrocorrinate complex is obtained, while the analogous β-unsubstituted species is fully reduced to the corresponding corrinate.

5.1.2 Syntheses of Corrins

The method involving hydrogenation of a β-unsubstituted octadehydrocorrinate represents an easy approach to the synthesis of corrins, but it is not applicable to the preparation of natural compounds, which have very complicate substitution patterns with nine asymmetric centres.

The first synthesis of a corrin reported in the literature utilized a linear tetrapyrrole containing the direct link between the A and D pyrroles [64, 65]. Such an elegant although lengthy approach differs from those outlined above where the formation of the direct A–D link represents the final step of the synthetic procedure.

Eschenmoser conceived a synthesis of corrins which could be extended to biological compounds. In this method a Cd^{2+} complex of the secocorrin which formula is shown in Fig. 24a ($R = {=}CH_2$, $R' = H$, $R'' = CN$) is converted

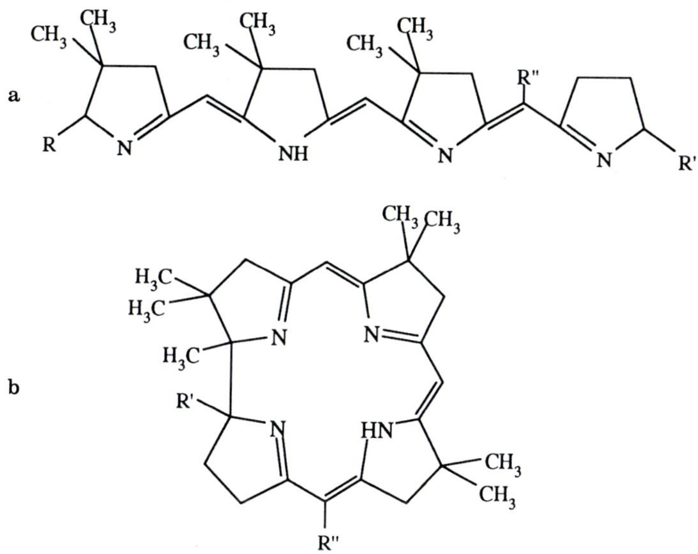

Fig. 24a, b. The structural formula of the secocorrin **a**) utilized for the syntheses of corrins **b**) (See Sect. 5.1.2)

into the corresponding corrin photochemically, in agreement with the Woodward–Hoffmann rules.

Such photochemical cyclization may be also accomplished when the secocorrin is coordinated to Li^+, Mg^{2+}, Zn^{2+}, Pd^{2+} or Pt^{2+}, but fails with Cu^{2+}, Ni^{2+}, Co^{2+} or Mn^{2+}. The metal-free ligand is also unable to cyclize [66, 67].

A direct synthesis of a Ni complex of corrin may be carried out thermally when the secocorrin shown in Fig. 24 a ($R = CH_3$, R' = formyl or acetyl, R'' = CN) is heated in the presence of Ni salts [68, 69]. The corresponding 1-methyl-19-unsubstituted corrinate Ni^{2+} is then achieved by base-induced removal of the formyl or acetyl group.

Eschenmoser has also described an electrolytic, reductive, acid catalyzed cyclization of the Ni complex of 18,19-dehydrosecocorrin [67]. Such a reaction has been considered a model for the biosynthesis of corrinoid rings.

The formation of corrin complexes with metals different from Ni or Co has been reported by Koppenhagen et al. [70]. However these corrinates are strictly connected to Vitamin B_{12} and will not be discussed further.

5.1.3 Syntheses of Di-, Tetra- and Hexa-dehydrocorrins

Corrin and octadehydrocorrin represent the extremes in the family of corrinoid compounds: between these two structures there are those macrocycles in which additional double bonds can be systematically introduced in the ring.

These compounds have been obtained by cyclization of the appropriate reduced bilin but the resulting conversions operate non-oxidatively and are acid catalyzed.

The substituent pattern is that shown in Fig. 24b unless otherwise indicated, the structural differences are in the level of unsaturation present in the macroring.

D-Didehydrocorrin was obtained for the first time as its Ni complex, D-Ni(ADDC), by Eschenmoser's group [71, 72]. Also, the reduction of 8,12-diethyl-1,2,3,7,13,17,18,19-octamethyl-octadehydrocorrinato nickel chloride has been reported to afford the corresponding D-didehydrocorrinate complex [52–54] (see Sect. 5.1.1).

In 1982, Kräutler and Hilpert reported on the preparation of (D-didehydrocorrinate)(dicyano)Co(III) by non-oxidative photocyclization of 19-cyanosecocorrin (Fig. 24a: (R = $=CH_2$, $R' = R'' = CN$) to the corresponding 1-methyl-19-cyanocorrinate Cd^{2+} [73]. Demetallation, subsequent insertion of cobalt and careful thermal elimination of the 19-cyano group affords the didehydrocorrinato cobalt complex as illustrated in Fig. 25. The complex has been converted into the corresponding corrinate by treatment with zinc in acetic acid.

C,D-Tetradehydrocorrins have been prepared during the studies on the synthesis of isobacteriochlorins [74]. They represent the only example of reduced corrins isolated as metal-free ligands.

The synthetic pathways leading to tetradehydrocorrins and isobacteriochlorins are very similar and it is just by fine variations of the reaction conditions that the preparation is driven towards specific tetrapyrrolic rings. The linear precursors have been synthesized using the sulfide contraction method (also indicated by other authors as "sulfur extrusion"). The corrin skeleton is formed by alkaline hydrolysis of the cyano protecting group present at the 19 position and subsequent acid catalyzed coupling of pyrroles A and D, as described in Fig. 26.

Another method used for the synthesis of C, D-Ni(ATDC) has been the anaerobic cyclization of nonamethyl bilinogen [75] illustrated in Fig. 27. This linear tetrapyrrole has been obtained by reduction of the corresponding dihydrobilin dihydrobromide salt with $NaBH_4$ in 80% aqueous methanol.

The reaction affords a complex mixture of products containing the diastereoisomers of a Ni(II) C,D-tetradehydrocorrinate. In addition to this major product, other nickel corrinoids have been isolated from the reaction mixture containing: Ni(AODC), B,C,D-Ni(AXDC) and A,C,D-Ni(AXDC) which surprisingly are derived from dehydrogenation of the parent complex, despite the anaerobic conditions of the reaction.

The nature of the products has been demonstrated to depend dramatically on the reaction conditions: the use of an organic base, 1,5,7-triazabicyclo-[4.4.0]dec-5-ene, is necessary, because other bases, such as triethylamine, do not at all afford the complex. Other small variations, such as the use of larger quantities of the base, lead to unexpected products, such as tetrahydroporphyrins.

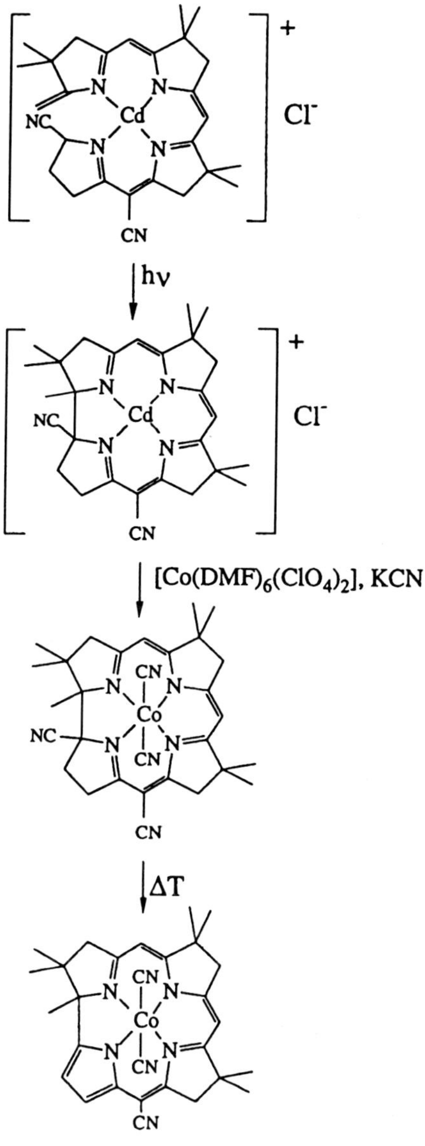

Fig. 25. Reaction sequence leading to the formation of (D-didehydrocorrinato)(dicyano)-cobalt(III). According to Ref. [73]

B,C,D-Ni(AXDC) fills the last gap in the corrin family and its preparation has been reported by Monforts [76, 77]. This compound has been obtained by cyclization of a Ni complex of the linear tetrapyrrole reported in Fig. 28. The synthetic pathway is similar to that reported for tetradehydrocorrins: a cyano group is utilized to protect the 1-position and then removed by alkaline hydrolysis. Cyclization may be acid catalyzed at room temperature, or accomplished thermally at 240–260 °C, with similar yields.

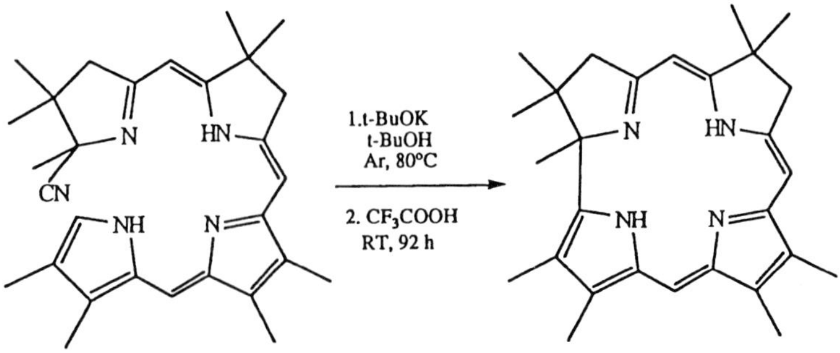

Fig. 26. Synthesis of C,D-tetradehydrocorrin. According to Ref. [74]

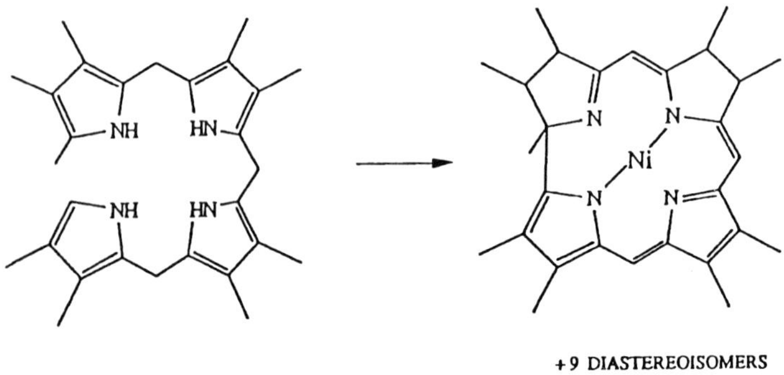

Fig. 27. Anaerobic cyclization of nonamethylbilinogen to Ni(ATDC). According to Ref. [75]

5.2 Syntheses of Corrinoids via Ring Contraction Reactions

The synthetic procedure involving ring contraction of hydroporphyrins has been applied only to the preparation of corrin complexes [78] and it has been inspired by the studies on the biosynthesis of Vitamin B_{12}. The macroring precursor is the hexahydroporphyrin shown in Fig. 29.

The substituents present at the 20 position are characteristic of this structure: the methyl group is present in Factor III, an intermediate in the biological conversion of uroporphyrinogen to Vitamin B_{12}. The hydroxyl group represents the hypothetical subsequent biological transformation. Their presence is essential for ring contraction: in the absence of the 20-methyl group the conversion to the corrin skeleton does not occur, but there is tautomerizaton to a keto form [69, 79–81].

Ring contraction to 1-methyl-19-acetylcorrinate Ni^{2+}, or to the analogous (1-methyl-19-acetylcorrinate)(dicyano)Co^{3+}, occurs when the corresponding hexahydroporphyrinoid complex is heated for a few minutes above its melting point under anaerobic conditions [78]. Base induced elimination of the acetoxy angular substituent affords the 1-methylcorrin complex, which has a structure similar to that of its biological analog.

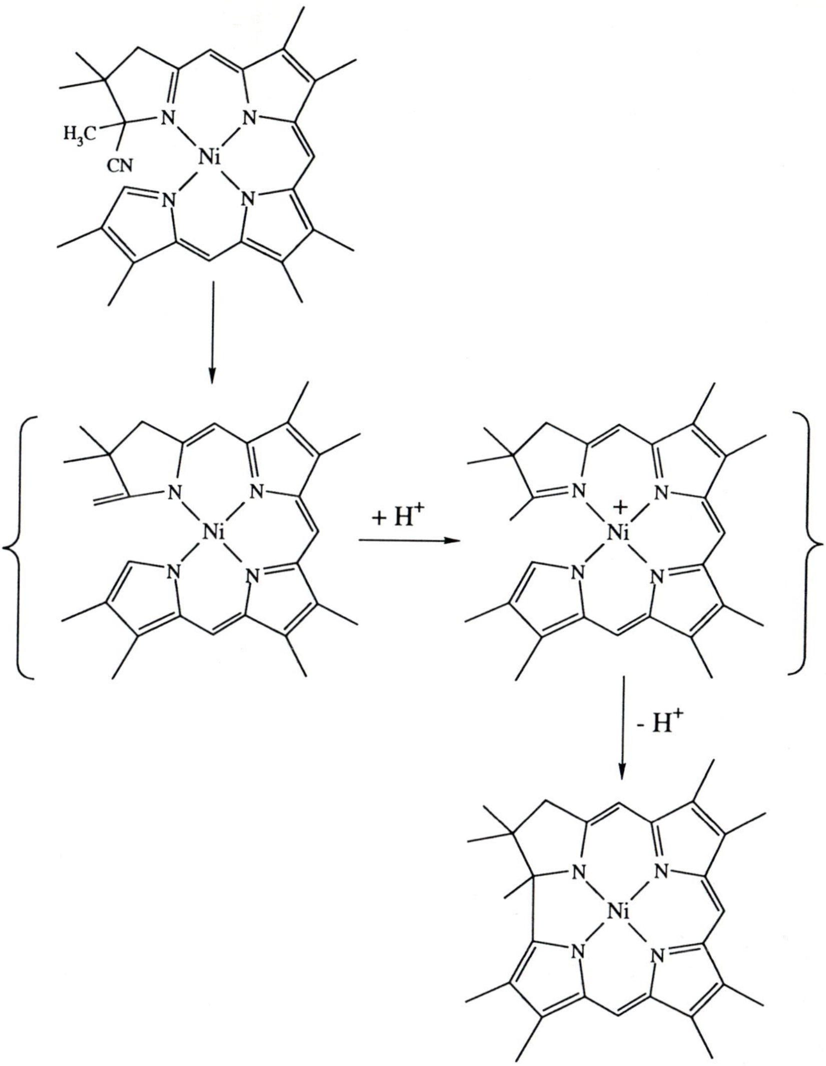

Fig. 28. Ring closure reaction leading to Ni(AXDC). According to Ref. [77]

The synthetic conditions are unsatisfactory with respect to the intended role of the reaction as a potentially biomimetic model, but demonstrate that it is possible to achieve interconversion between the structures of hydroporphyrin, secocorrin and corrin.

Coordination of Ni or Co is necessary to achieve ring contraction: in fact the metal free hydroporphyrin, or its Zn complex, does not afford corrin, but ring opening occurs to give the corresponding secocorrin complex [69]. This seems to be a general feature: hexahydroporphyrinoid ligands prefer corrin like

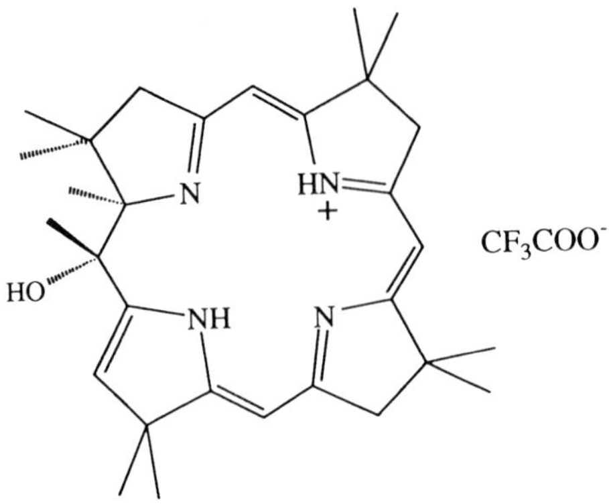

Fig. 29. The structural formula of the hexahydroporphyrinoid precursor of the corrin structure

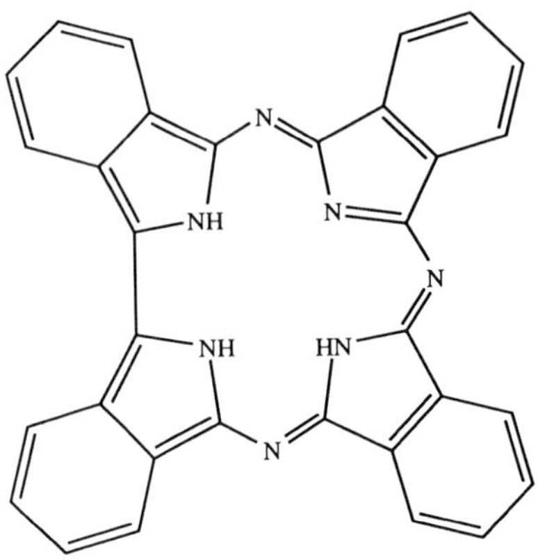

Fig. 30. The structural formula of triazatetrabenzcorrole, H_3(TBC)

arrangements in the presence of suitable metal ions, otherwise they preferentially exist as hydroporphyrins.

The size of the macrocycle coordination "hole" has a profound influence on its coordination chemistry: thus in the weak ligand field of hydroporphyrinoid systems Ni^{2+} assumes the high spin state and binds extra axial ligands, while in the stronger ligand field of the corrin macrocycle the coordination sphere of the metal is saturated and diamagnetic complexes are generated [82].

Another example of a ring contraction reaction leading to a corrinoid structure is the preparation of α,β,γ-triazatetrabenzcorrole [83] the formula of which is reported in Fig. 30. This new macrocyclic ring has been isolated as its Ge^{4+} derivative formed by the ring-contractive reaction of germanium phthalocyanine carried out in the presence of $NaBH_4$ or H_2Se.

Si, Al and Ga derivatives have also been prepared but they could not be fully characterized because they are highly sensitive to moisture and oxygen and readily decompose.

Table 19. Synthetic metal complexes of corrinoid macrocycles

Macrocycle[a]	Metal	Reference
$(A_2ODC)^-$	Ni	51, 52, 54, 57
	Co	2, 51, 52
$(AODC)^{2-}$	Co	54, 56, 86
$(C_2ODC)^-$	Ni	51, 60
	Co	51, 60
	Pd	60
	Pt	61
$(ACODC)^-$	Ni	51
$(A_2XDC)^-$	Ni	53
$(AXDC)^{2-}$	Ni	54, 76, 77
$(A_2TDC)^-$	Ni	54
	Co	2
$(ATDC)^{2-}$	Ni	75
$(A_2DDC)^-$	Ni	51, 52, 54
$(ADDC)^{2-}$	Ni	72
	Co	73
$(TBC)^{3-}$	Ge	83
	Si	83
	Al	83
	Ga	83

[a] See list of abbreviations (Sect. 2.3)

The electronic spectra of the complexes are characterized by strong absorptions at 440–450 and 650–680 nm and show only slight dependence on the central metalloid. As for other corrinoids, the Q and Soret bands arise from a combination of several vibrational components of transitions within the π system. The molecular symmetry of the complex has been identified as a two-fold symmetry with some distortion.

The hydroxygermanium derivative resulted to be very light sensitive in solution giving rise to ring cleavage or ring expansion reactions which have not been fully characterized.

A summary of the corrinoid complexes synthesized up to date is reported in Table 19.

6 Structural Properties and Reactivity of Corrinoids

The synthetic procedures leading to corrinoid metal complexes have been developed in order to use them as models of the biosynthetic pathway to Vitamin B_{12}. Their chemical and structural properties were expected to be very similar to those of the natural coenzyme and this is probably the reason why very few detailed investigations on their spectroscopic or electrochemical features and their reactivity towards axial coordination have been carried out.

The only species that have been the subject of extensive studies are the metal complexes of octadehydrocorrins and these investigations have been reviewed in the past [8, 10, 11, 59].

Further reports have been published by Murakami and co-workers: they focused their attention on the cobalt complexes of octadehydrocorrin and

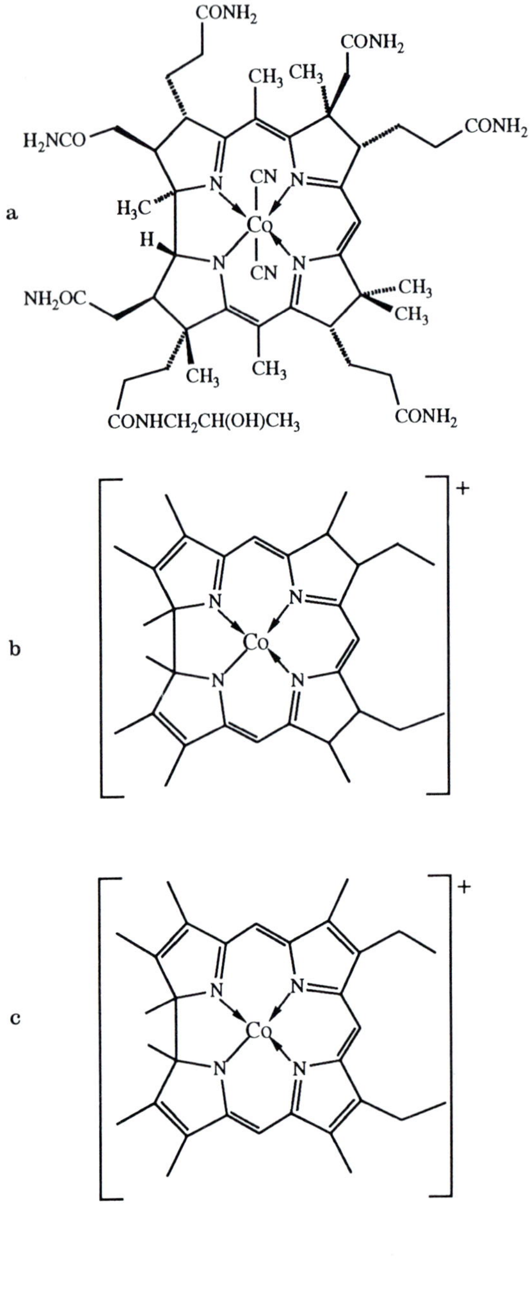

Fig. 31a–c. The structural formulae of the cobalt complexes of **a**) cobinamide, **b**) 1-19-dimethyl-A,D-tetradehydrocorrin, and **c**) 1,19-dimethyloctadehydrocorrin

tetradehydrocorrin with respect to the chemistry of the naturally occurring corrinoids [2]. Corrin has been studied as an equatorial ligand which enables the cobalt ion to carry out several otherwise unusual reactions. Thus synthetic corrinoids where minimal variations have been performed on the macrocyclic skeleton have been investigated. The structural formulae of such compounds are shown in Fig. 29 together with that of the parent system, cobinamide (Fig. 31a).

Additional double bonds have been inserted at the peripheral positions of the macroring: in tetradehydrocorrin (A_2TDC)- (Fig. 31b) unsaturations are present at A and D pyrroles, while it is the β positions of octadehydrocorrin (A_2ODC)- (Fig. 31c) which are fully unsaturated. Murakami has used the name bidehydrocorrin for tetradehydrocorrin and tetradehydrocorrin for octadehydrocorrin. In order to avoid confusion the latter names will be used here as described in Sect. 2.

Another structural variation operated in these compounds is the substitution at the angular 1,19-positions with methyl groups: in natural corrinoids only the 1 postion is substituted. Such alteration is necessary on one hand because the cobalt complex of 1-methyl-octadehydrocorrin is unstable and, on the other hand, the related $[Co(A_2TDC)]^+$ is prepared by reduction of $[Co(A_2ODC)]^+$ [2].

Although the basic structure of these synthetic species is very similar to that of the natural corrinoids their reactivity shows several differences; thus $[Co(A_2ODC)]^+$ does not form an adduct with dioxygen or give organometallic derivatives, which are characteristic features of the chemistry of Vitamin B_{12} and related compounds [13]. This failure has been attributed to a different electronic structure, as confirmed by the very different optical spectra of the two systems (Fig. 32).

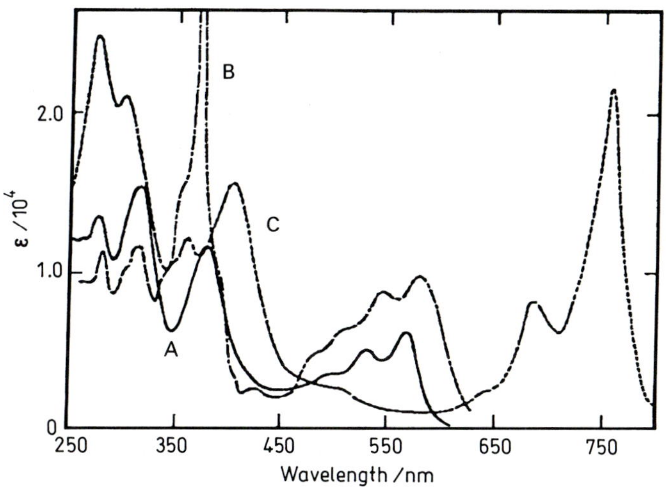

Fig. 32. The electronic absorption spectra of *A*) $Co(CN)_2(A_2TDC)$ in water, *B*) $Co(CN)_2(Cbi)$ in water, *C*) $Co(CN)_2(A_2ODC)$ in methanol. Taken from Ref. [2] with permission

The spectral indications are also confirmed by cyclic voltammograms: in Table 17 redox potentials of $[Co(A_2ODC)]^+$ and $[Co(A_2TDC)]^+$ are reported, compared to those of other corrinoids. Redox pairs are all metal centered; $[Co(A_2ODC)]^+$ presents an extra wave in aprotic solvents attributable to the $[Co(A_2ODC)]/[Co(A_2ODC)]^-$ process. The dicyano-complexes show only one redox couple, related to a two electrons reduction of Co^{3+} to Co^+. A peculiar feature of $[Co(A_2ODC)]^+$ with respect to other corrinoids is the significant shift of the Co^{2+}/Co^+ potential which indicates a very high stability of the lower oxidation state of cobalt in this complex. Such stability may explain the lack of formation of adducts with dioxygen, or of organometallic derivatives when the complex is treated with alkyl halides.

The stronger diamagnetic shielding of the resonances of the *meso*-protons in the NMR spectrum of $[Co(CN)_2(A_2ODC)]$ with respect to the corresponding tetradehydrocorrin complex is consistent with the existence of an extended π conjugation.

Among other effects, such as an extended π conjugation may stabilize the Co^+ state accepting an electron from the metal into the lower π^* orbital. Upon reduction of $Co(A_2ODC)$ two electrons are transferred to the π^* level of the ligand, leading to $[Co(A_2ODC)]^-$, where the metal atom is formally in the bivalent state.

The nature of the additional double bonds depends on their locations on the macroring: unsaturations present at B and C pyrroles are conjugated to the inner double bonds and are responsible to the extension of the π system, while those present at A and D rings are of isolated character; $(A_2TDC)^-$ has in fact ligating properties similar to those of natural corrinoids and may thus be considered a good model compound.

Table 20. Half wave potentials/V vs SCE of cobalt corrinoid complexes. According to Ref. [2]

Complex[a]	Solvent	Co^{3+}/Co^{2+}	Co^{2+}/Co^+	$[Co^+/Co^{2+}(L)]^-$ [b]	$(CN)_2Co^{3+}/Co^+$
$[Co(A_2TDC)]^+$	CH_3OH	0.47	−0.71		
	DMF		−0.55		
	CH_2Cl_2		−0.57		
$[Co(A_2ODC)]^+$	CH_3OH	0.59	−0.25		
	DMF		−0.11	−1.31	
	CH_2Cl_2	0.97	−0.13	−1.57	
Vitamin B_{12}	H_2O	0.3	−0.742		
$Co(CN)_2(A_2ODC)$	DMF				−0.58
	CH_2Cl_2				−0.73
$Co(CN)_2(A_2TDC)$	DMF				−1.38
	CH_2Cl_2				<-1.45
$Co(CN)_2(Cbi)$	DMF				−1.14
	H_2O				−1.18

[a] See list of abbreviations (Sect 2.3).
[b] L indicates the macroring. $[Co^+/Co^{2+}(L)]^-$ is the redox potential for $Co^+(A_2ODC) \rightleftarrows [Co^{2+}(A_2ODC)]^-$.

Thus $[Co(A_2TDC)]^+$ forms a dioxygen adduct and organometallic derivatives. Closely related electronic structures are confirmed by spectral data, such as EPR, stability constants and CV, which are reported in Tables 20, 21 and 22.

Nature has furnished the corrin ring of a conjugation system which enables the cobalt atom to carry on metabolic reactions: tetradehydrocorrin, having a similar π system, is therefore a better model than octadehydrocorrin.

However, $[Co(A_2TDC)]^+$ fails to catalyze 1,2-rearrangements, important physiological reactions induced by corrinoid-dependent enzymes. This failure has been attributed to the different substitution pattern at the 1,19 angular positions of the ring.

Having one angular position unsubstituted natural corrinoids have one side of the ring accessible to interactions with axial substituents with no steric constraints. This is not the case of $[Co(A_2TDC)]^+$ where steric interactions may prevent an adequate approach of the complex to the substrate. When reacted with halogenated β-diesters, such as dimethyl bromomethylmalonate, instead of rearrangement, $[Co(A_2TDC)]^+$ catalyzes reductive dehalogenation affording dimethyl methylmalonate. Although its presence has not been evidenced the formation of a metastable alkylated complex has been postulated to occur during the reaction since no dimethyl methylmalonate has been detected in the absence of the complex.

Table 21. EPR Spin Hamiltonian parameters of Co^{2+} corrinoid complexes at 77 K. According to Ref. [2]

Complex	Medium	$10^4\ A_{\parallel}^{Co}/cm^{-1}$	$10^4\ A_{\parallel}^{N}/cm^{-1}$
$[Co(A_2TDC)]^+$	$CHCl_3/C_6H_6$ (2:1 v/v)	127	
$[Co(A_2TDC)Py]^+$	$CHCl_3/C_6H_6$ (2:1 v/v)	101	17.1
$[CoO_2(A_2TDC)]^+$	$CHCl_3/C_6H_6$ (2:1 v/v)	18	
$[CoO_2(CTDC)Py]^+$	$CHCl_3/C_6H_6$ (2:1 v/v)	12	
$[Co(Cbi)]^+$	H_2O	135	
$[Co(Cbi)Py]^+$	CH_3OH	105	17.7
Vitamin B_{12} r	H_2O	102.7	
Vitamin B_{12} r O_2	H_2O	14	

Table 22. Stability constants concerning to the coordination of pyridine bases to $Co(A_2TDC)ClO_4$ in dichloromethane. $(A_2TDC)^-$ represents the anion of 8,12-diethyl-1,2,3,7,8,12,13,17,18,19-octamethyl-tetradehydrocorrin. According to Ref. [2]

Base	pK_a	K $(l\,mol^{-1})$
Pyridine	5.19	$2.7 \cdot 10^3$
4-Methylpyridine	6.02	$5.1 \cdot 10^3$
4-Aminopyridine	9.12	$6.1 \cdot 10^3$
4-Cyanopyridine	1.90	$2.1 \cdot 10^2$
2-Methylpyridine	5.97	9.1
2,4,6-Trimethylpyridine	7.48	8.4

Steric repulsions also influence the photolytic cleavage of the cobalt–carbon bond in organometallic complexes: with methyl and ethyl derivatives the process is homolytic and affords a Co^{2+} complex, whilst when bulkier alkyl ligands, such as an isopropyl group, are present a different pathway is followed and photolysis affords $[Co(A_2TDC)]^{2+}$. The reaction rate is unaffected by the presence of oxygen, which may act as a radical scavenger, and this feature, coupled with the formation of a Co^{3+} complex, strongly indicates an heterolytic bond cleavage.

This novel bond-breaking pattern may be attributed to the significant steric strains involved in the isopropyl derivative which may be the origin of the instability of Co-organoderivatives of $(A_2TDC)^-$, which in some case decompose even in anaerobic conditions in the dark.

The various oxidation states accessible to cobalt and their possible interconversion in these complexes have also been studied [84, 85].

$[Co(A_2ODC)]^{2+}$ has been obtained by oxidation of $[Co(A_2ODC)]^+$ (Fig. 31c) in acidic media [84]. The reaction rate is not affected by the presence of axial ligands, such as cyanide, and the final product is the diaquaderivative $[Co(H_2O)_2(A_2ODC)]^{2+}$ with no dependence on the nature of the acid used.

Acid dissociation constants for the aqua ligands have been measured and their values are comparable to those of other corrinoids. Figure 33 shows the electronic absorption spectra of the species present at different pH values.

$[Co(A_2ODC)]^{2+}$ is easily reduced to the Co^{2+} complex in aqueous carbonate solution (1% v/v methanol). This reaction is photocatalyzed and induced by

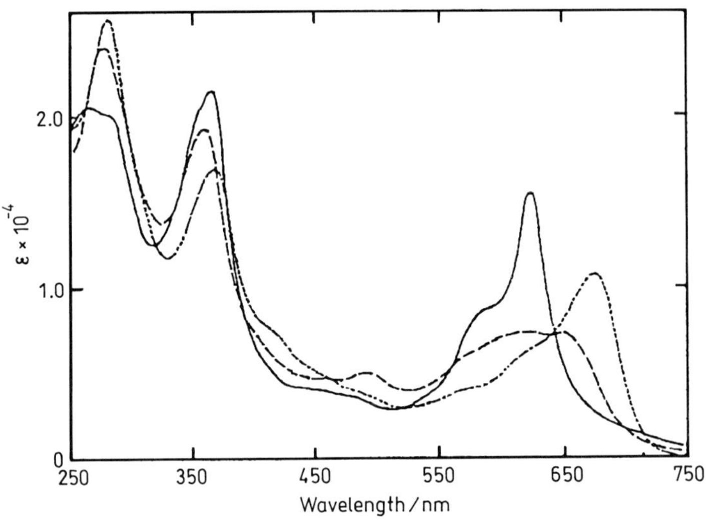

Fig. 33. The electronic absorption spectra of diaqua [——] (pH < 6.5), aquahydroxo [– – –] (6.5 < pH < 11.5) and dihydroxo [· · · · ·] (pH > 11.5) species derived from $[Co(A_2ODC)]^{2+}$ in water/methanol (99.2:0.8 v/v) at 10 °C. Taken from Ref. [84] with permission

carbonate salts: phosphate or borate ions as buffer salts are less effective at promoting the reduction. Kinetic analysis of the reaction indicated the following as a plausible mechanism:

$$Co^{III}OH \rightleftarrows Co^{II} + {}^{\bullet}OH$$

$$^{\bullet}OH + CO_3^{2-} \rightarrow OH^- + CO_3^{-\bullet}$$

$$CO_3^{-\bullet} \rightarrow \text{further reaction}$$

The first step is an homolytic cleavage of the Co–OH bond with the formation of the hydroxyl radical which is then scavenged by the carbonate ion.

When an organic co-solvent, such as acetonitrile or *t*-butyl alcohol, is added the reaction rate increases. This effect has been attributed to a variation in the solvation status of the reacting complex: a lower solvation of the hydroxo complex may facilitate the formation of a neutral hydroxo radical. A radical scavenging of these organic solvents has been excluded because the presence of an excess of carbonate ions assures that the rate-determining step is the homolytic cleavage of the Co–OH bond.

The reduction has also been carried out by adjusting the pH value with hydroxide ions. The reaction must now follow a different pathway: the reduction is still photocatalyzed, but at pH > 13 a sharp decrease of the rate has been observed. The authors proposed the mechanism outlined below:

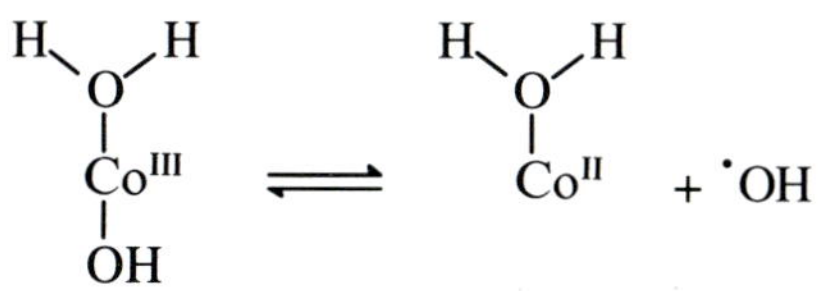

$$^{\bullet}OH \underset{-OH^-}{\overset{+OH^-}{\rightleftharpoons}} O^{-\bullet} + H_2O$$

Since the aquohydroxo complex is the active species the decrease of the reaction rate at pH > 13 has been explained with the decrease of the concentration of this reacting form. The first step of the reaction is the homolytic cleavage of Co–OH bond, followed by deprotonation of the resulting hydroxo radical with formation of $O^{-\bullet}$, which eventually dimerizes to give hydrogen peroxide, as revealed by iodometry. Deprotonation of the hydroxo radical is necessary to make less effective the backward reaction of recombination with the Co^{2+} complex.

A different redox behavior has been observed in strongly alkaline aqueous media ($[OH^-] > 1$ M) [85]. $Co(A_2ODC)^{2+}$ is reported to be reduced to the well characterized Co^+ species [86]. This reduction is strongly dependent on the hydroxide concentration and it is not influenced by the presence of carbonate ions or other radical scavengers. A homolytic cleavage of the Co–OH bond has then been excluded. Again kinetic data have been used to propose the following mechanism:

$$[Co^{III}]_2 \rightleftarrows 2Co^{III}$$

$$Co^{III} + OH^- \rightleftarrows [Co^{III}OH]^- \leftrightarrow [Co^{I}OH]^+$$

$$\{[Co^{III}OH]^- \leftrightarrow [Co^{I}OH]^+\} + OH^- \rightarrow Co^{I} + H_2O_2$$

The initial presence of a dimeric species may explain the dependence of the rate on the square root of the concentration of $[Co(A_2ODC)]^{2+}$. Subsequent heterolytic cleavage of the Co–OH bond affords the Co^+ complex and hydrogen peroxide.

The same reduction has also been carried out at lower pH values using a thiol as reductant, such as $HS(CH_2)_2OH$.

$[Co(A_2ODC)]^+$ disproportionates to an equimolar mixture of Co^+ and Co^{3+} species when reacted in similar conditions ($[OH^-] > 1$ M) [85]. The electronic spectrum of the reaction mixture results from the superposition of the absorptions of the two complexes and is identical to a calculated spectrum, as shown in Fig. 34. At lower hydroxide concentration the disproportionation is incomplete and an equilibrium is present:

$$2\,Co^{II} + 2\,OH^- \rightleftarrows [Co^{III}(OH)_2]^+ + Co^+$$

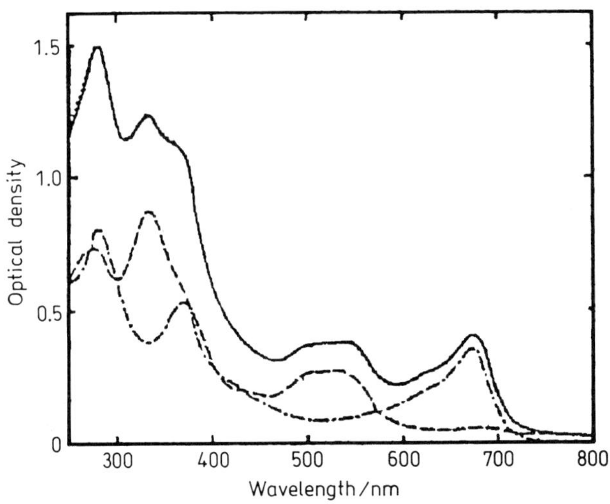

Fig. 34. The electronic absorption spectra of cobalt complexes of octadehydrocorrin in water/methanol (99:1 v/v) at 3 °C: —— $[Co(A_2ODC)]^+$, 6.38×10^{-5} M after disproportionation at $[OH^-] = 1.0$ M; – · – · – $Co(OH)_2(A_2ODC)$, 3.19×10^{-5} M at $[OH^-] = 0.2$ M; – – – Co(AODC), 3.19 $\times 10^{-5}$ M at $[OH^-] = 0.1$ M; · · · · · calculated spectrum for a solution containing $Co(OH)_2(A_2ODC)$, 3.19×10^{-5} M and $Co(A_2ODC)$, 3.19×10^{-5} M. Taken from Ref. [85] with permission

The equilibrium constant has been spectrophotometrically determined and its average value resulted to be K = 552.

Very recently a study on nickel corrinoids has been reported [87] where the nature of the product of the aerobic oxidation of (1,2,3,7,13,17,18,19-octamethyl-8,12-diethyl-octadehydrocorrinato)nickel(II) chloride has been elucidated by means of X-ray crystallography.

At variance with the formulation reported by Johnson et al. over 20 years ago [7] who postulated the formation of an epoxychlorin mainly on the basis of optical spectroscopy, the complex resulted to be an unexpected tetraaza macrocycle consisting of only three pyrrole rings, with the fourth one reassembled into an exocyclic furan ring (Fig. 35a).

The corrinoid macrocycle has undergone a complicated series of reactions in order to give the final product: the corrin ring has expanded, one angular methyl

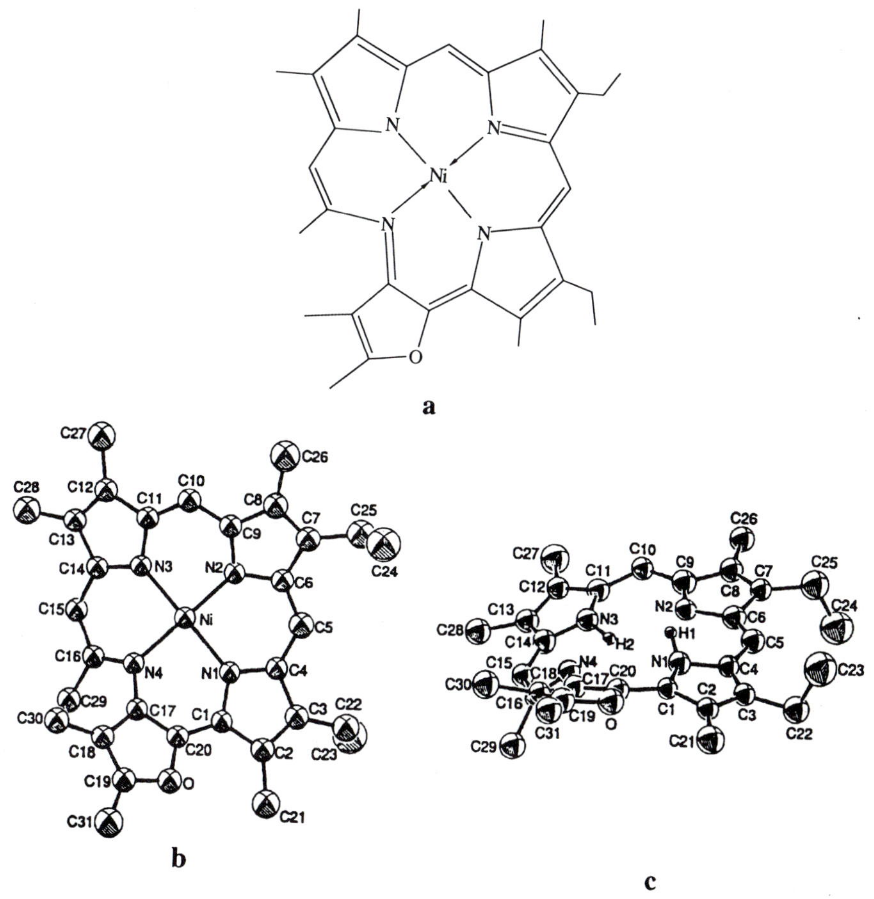

Fig. 35a–c. The structural formula of the nickel complex of furochlorophin (**a**), its crystal structure (**b**) and the crystal structure of the free base (**c**). Taken from Ref. [87] with permission

group has become a methine bridge, a pyrrole ring has opened and the vinyl group created has reattached itself to the macrocycle via an oxide. No detailed mechanism has been proposed for such a complex pathway.

The final product, for which the authors have proposed the name furochlorophin, has an 18 π electron aromatic system. The nickel atom is coplanar with the four nitrogen atoms but the molecule deviates up to 1.93 Å from planarity (Fig. 35b).

Nickel furochlorophin can be demetallated by treatment with concentrated sulfuric acid giving in good yield (60%) the free base the structure of which has also been resolved by X-ray crystallography (Fig. 35c). The macrocycle is more planar than its nickel complex (maximum deviation about 1 Å) and can be easily converted into its Cu^{2+} and Zn^{2+} derivatives.

Very recently [88] a synthetic derivative of Vitamin B_{12} where five viologen units have been covalently linked to the β side chains of the corrin nucleus has been reported in the literature (Fig. 36).

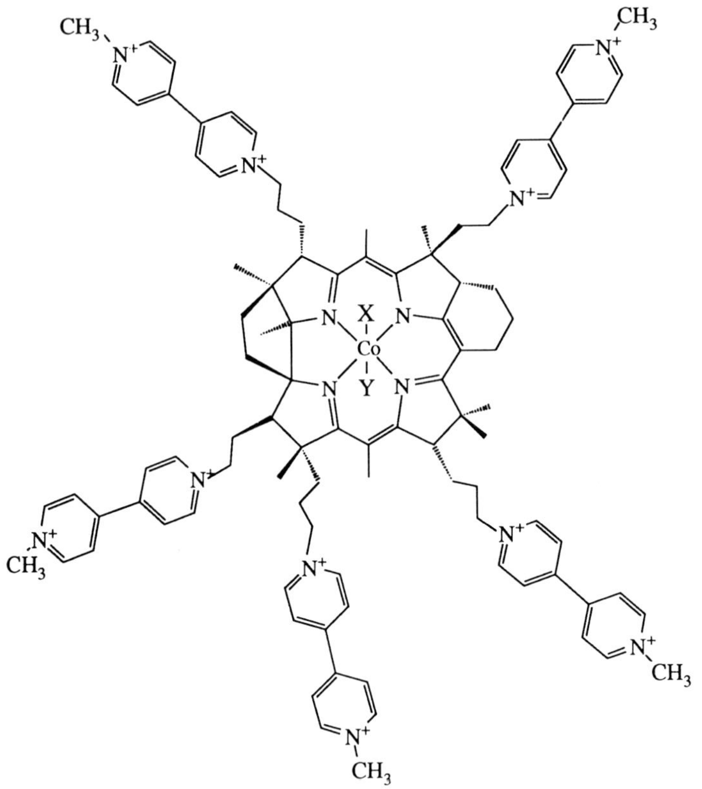

Fig. 36. The structural formula of the (pentaviologen-corrinato)cobalt(II) complex. X or Y = ClO_4^-; Y or X = H_2O. According to Ref. [88]

The redox activity of the cobalt atom coordinated to such corrinoid structure resulted to be strongly influenced by the presence of the five viologen groups as revealed by cyclic voltammetry and spectroelectrochemistry. The new features observed have been attributed to the steric arrangement caused by the viologen units which do not allow a direct interaction of the cobalt atom with the electrode. The viologen units are involved only as bridges for electron transfer to or from the metal. The rate of the Co^{3+}/Co^{2+} redox reaction is markedly slowed down: redox waves attributable to such process are in fact absent in the voltammograms and only spectroelectrochemistry allows to demonstrate the reduction, which needs 30 min for equilibration because of a 710 mV activation barrier, needed for the viologen mediated electron transfer.

(Pentaviologen-corrinato)Co(II) has also been investigated as an electrocatalyst which might couple the well known redox properties of cobalt corrinoid complexes with a ten-electron storage system represented by the viologen units. The reaction studied has been the reductive conversion of 1,2-dibromocyclohexane to cyclohexene. After complete reduction of the complex to its Co(I) form by electrolysis, multiple reductive eliminations from dibromocyclohexane generating cyclohexene have been observed with concomitant reoxidation to the Co(III) form of the electrocatalyst, demonstrating an efficient electron transfer from viologens to cobalt. Unfortunately the viologen groups are very easily attacked by radicals so that the charge-storing ability of the complex is quickly lost if radicals are generated during the reaction. The activity of (pentaviologen-corrinato)Co(II) is thus limited to reactions which do not involve free radicals.

However, the electrochemical properties of such a complex are certainly very interesting since it represents a synthetic model for the activity of redox proteins and may mimic their capability to store reducing power in an oxidizing environment.

7 References

1. Stolzenberg AM, Stershic MT (1988) J Am Chem Soc 110: 6391 and Refs. therein
2. Murakami Y, Aoyama Y, Tokunaga K (1980) J Am Chem Soc 102: 6736
3. Johnson AW (1967) Chem in Britain 3: 253
4. Johnson AW, Kay IT (1965) J Chem Soc 1620
5. Battersby AR (1986) Acc Chem Res 19: 147
6. Harrison HR, Hodder OJR, Hodgkin DC (1971) J Chem Soc B 640
7. Johnson AW (1975) In: Smith KM (ed) Porphyrins and metalloporphyrins. Elsevier, Amsterdam, p 729
8. Grigg R (1978) In: Dolphin DH (ed) The porphyrins, vol II. Academic Press, NY
9. Jackson AH (1978) In: Dolphin DH (ed) The porphyrins, vol I. Academic Press, NY, p 341
10. Genokhova NS, Melent'eva TA, Berezowskii VM (1980) Russ Chem Rev 49: 1056
11. Melent'eva TA (1983) Russ Chem Rev 52: 641
12. Woodward RB (1973) Pure Appl Chem 33: 145
13. Golding BT (1979) In: Haslam E (ed) Comprehensive organic chemistry. Pergamon, Oxford, p 549
14. Fuhrer W, Schneider P, Schilling W, Wild H-J, Naag H, Obata N, Holmes A, Schreiber J, Eschenmoser A (1972) Chimia 26: 320

15. IUPAC-IUB Commission for Chemical Nomenclature (1974) Biochemistry 13: 1555
16. Bonnet R (1978) In: Dolphin DH (ed) The porphyrins, vol I. Academic Press, NY
17. Pandey RK, Zhou H, Gerzevske K, Smith KM (1992) J Chem Soc Chem Commun 183
18. Murakami Y, Aoyama Y, Hayashida M (1980) J Chem Soc Chem Commun 501
19. Grigg R, Johnson AW, Shelton G (1971) J Chem Soc (C) 2287
20. Adler AD, Longo FR, Finarelli JD, Goldmacher J, Assour J, Korsakoff L (1967) J Org Chem 32: 476
21. Boschi T, Licoccia S, Paolesse R, Tagliatesta P (1988) Inorg Chim Acta 141: 169
22. Boschi T, Licoccia S, Tagliatesta P (1986) Inorg Chim Acta 119: 191
23. Hanson LK, Gouterman M, Hanson JC (1973) J Am Chem Soc 95: 4822
24. Boschi T, Licoccia S, Paolesse R, Tagliatesta P, Tehran MA, Pelizzi G, Vitali F (1990) J Chem Soc Dalton Trans 463
25. Paolesse R, Licoccia S, Boschi T (1990) Inorg Chim Acta 178: 9
26. Buchler JW (1975) In: Smith KM (ed) Porphyrins and metalloporphyrins. Elsevier, Amsterdam, p 157
27. Vogel E, Will S, Schulze Tilling A, Neumann L, Lex J, Bill E, Trautwein AX, Wieghardt K (1994) Angew Chem 106: 771, Int Edn Engl 33: 731
28. Licoccia S, Paci M, Paolesse R, Boschi T (1991) J Chem Soc Dalton Trans 461
29. Boschi T, Licoccia S, Tagliatesta P (1987) Inorg Chim Acta 126: 157
30. Paolesse R, Licoccia S, Fanciullo M, Morgante E, Boschi T (1993) Inorg Chim Acta 203: 107
31. Paolesse R, Licoccia S, Bandoli G, Dolmella A, Boschi T (1994) Inorg Chem 33: 1171
32. Hitchcock PB, McLaughlin GM (1976) J Chem Soc Dalton Trans 1927
33. Matsuda Y, Yamada S, Murakami Y (1980) Inorg Chim Acta 44: L309
34. Conlon M, Johnson AW, Overend WR, Rajapaksa D, Elson CM (1973) J Chem Soc Perkin Trans I 2281
35. Buchler JW (1978) In: Dolphin DH (ed) The porphyrins, vol I. Academic Press, NY, p 389
36. Hush NS, Dyke JM, Williams ML, Woolsey IS (1974) J Chem Soc Dalton Trans 395
37. Brewer CT, Brewer GA (1988) Inorg Chim Acta 154: 67
38. Zanoni R, Boschi T, Licoccia S, Paolesse R, Tagliatesta P (1988) Inorg Chim Acta 145: 175
39. Macquet JP, Millard MM, Theophanides T (1978) J Am Chem Soc 100: 4741
40. Katz JJ, Brown CE (1983) Bull Magn Res 5: 3
41. Botulinski A, Buchler JW, Lee YJ, Scheidt WR, Wicholas M (1988) Inorg Chem 27: 927
42. Pawlik MJ, Miller PK, Sullivan EP, Levstik MA, Almond DA, Strauss SH (1988) J Am Chem Soc 110: 3007
43. Dugad LB, Mehdi OK, Mitra S (1987) Inorg Chem 26: 1741
44. Janson TR, Katz JJ (1978) In: Dolphin DH (ed) The porphyrins, vol IV. Academic Press, NY
45. Hush NS, Dyke JM, Williamson ML, Woolsey IS (1969) Molec Phys 17: 559
46. Barkigia KM, Berber MD, Fajer J, Medforth CJ, Renner MW, Smith KM (1990) J Am Chem Soc 112: 8851
47. Matsuda Y, Yamada S, Murakami Y (1981) Inorg Chem 20: 2239
48. Murakami Y, Matsuda Y, Yamada S (1981) J Chem Soc Dalton Trans 855
49. Kadish KM, Koh W, Tagliatesta P, Sazou D, Paolesse R, Licoccia S, Boschi T (1992) Inorg Chem 31: 2305
50. Hush NS, Woolsey IS (1974) J Chem Soc Dalton Trans 24
51. Dolphin D, Harris RLN, Huppatz JL, Johnson AW, Kay IT (1966) J Chem Soc C 30
52. Johnson AW, Overend WR (1972) J Chem Soc Perkin Trans I 2681
53. Johnson AW, Overend WR, Hamilton AL (1973) J Chem Soc Perkin Trans I 991
54. Dicker ID, Grigg R, Johnson AW, Pinnock H, Richardson K, van den Broek P (1971) J Chem Soc C 536
55. Inhoffen HH, Buchler JW, Puppe L, Rohbock K (1971) Liebigs Ann Chem 747: 133
56. Clarke DA, Grigg R, Harris RNL, Johnson AW, Kay IT, Shelton KW (1967) J Chem Soc C 1648
57. Grigg R, Johnson AW, Shelton KW (1968) J Chem Soc C 1291
58. Johnson AW (1975) Chem Soc Rev 4: 1
59. Johnson AW (1980) Chem Soc Rev 9: 125
60. Gossauer A, Maschler H, Inhoffen HH (1974) Tetrahedron Lett 1277
61. Engel J, Inhoffen HH (1977) Liebigs Ann Chem 767
62. Jeyakumar D, Snow KM, Smith KM (1988) J Am Chem Soc 110: 8562 and references therein
63. Boschi T, Paolesse R, Tagliatesta P (1990) Inorg Chim Acta 168: 83
64. Bertele E, Boos H, Dunitz JD, Elsinger F, Eschenmoser A, Felner I, Gribi HP, Gschwend H, Meyer EF, Pesaro M, Scheffold R (1964) Angew Chem 76: 393

65. Bertele E, Eschenmoser A, Gschwend H, Pesaro M, Scheffold R (1965) Proc Roy Soc London, Ser A 288: 306
66. Yamada Y, Miljkovic D, Wehrli P, Golding B, Löliger P, Keese R, Müller K, Eschenmoser A (1969) Angew Chem 81: 301
67. Eschenmoser A (1976) Chem Soc Rev 5: 377
68. Pfaltz A, Bühler N, Neier R, Hirai K, Eschenmoser A (1977) Helv Chim Acta 60: 2653
69. Rasetti V, Hilpert K, Fässler A, Pfaltz A, Eschenmoser A (1981) Angew Chem 93: 1108, Int Ed Engl 20: 1058
70. Koppenhagen VB, Ernst L, Grothjan L, Dresow B (1982) Liebigs Ann Chem 1575
71. Rasetti V, Kräutler A, Pflatz A, Eschenmoser A (1977) Angew Chem 89: 475
72. Kräutler B, Pfaltz A, Nordmann R, Hodgson KO, Dunitz JD, Eschenmoser A (1976) Helv Chim Acta 59: 924
73. Kräutler B, Hilpert K (1982) Angew Chem 94: 139
74. Ofner S, Rasetti V, Zehnder B and Eschenmoser A (1981) Helv Chim Acta 64: 1431
75. Angst C, Kratky C, Eschenmoser A (1981) Angew Chem 93: 275, Int Edn Engl 20: 263
76. Monforts FP (1982) Angew Chem 94: 208, Int Edn Engl 21: 214
77. Monforts FP, Bats JW (1987) Helv Chim Acta 70: 402
78. Rasetti V, Pfaltz A, Kratky C, Eschenmoser A (1981) Proc Natl Acad Sci USA 78: 16
79. Battersby AR, Matcham GWJ, McDonald E, Neier R, Thompson M, Woggon WD, Bykhovskii VYa, Morris HR (1979) J Chem Soc Chem Commun 185
80. Lewis NG, Neier R, Matcham GWJ, McDonald E, Battersby AR (1979) J Chem Soc Chem Commun 541
81. Müller G, Gneuss RD, Kriemler HP, Scott AI, Irwin AJ (1979) J Am Chem Soc 101: 3655
82. Kräutler B (1987) Chimia 41: 277
83. Fujiki M, Tabei H, Isa K (1986) J Am Chem Soc 108: 1532
84. Murakami Y, Aoyama Y, Tada T (1981) Bull Chem Soc Jpn 54: 2302
85. Murakami Y, Aoyama Y, Tada T (1981) Inorg Chim Acta 54: L111
86. Elson CM, Hamilton A, Johnson AW (1973) J Chem Soc Perkin I 775
87. Chang CK, Wu W, Chern SS, Peng SM (1992) Angew Chem 104: 61, Int Edn Engl 31: 70
88. Steiger B, Walder L (1992) Helv Chim Acta 75: 90

Photochemistry of Tetrapyrrole Complexes

J. Šima

Department of Inorganic Chemistry, Slovak Technical University, Radlinského 9,
812 37 Bratislava, Slovakia

The present article reviews the photochemical deactivation modes and properties of electronically excited metallotetrapyrroles. Of the wide variety of complexes possessing a tetrapyrrole ligand and their highly structured systems, the subject of this survey is mainly synthetic complexes of porphyrins, chlorins, corrins, phthalocyanines, and naphthalocyanines. All known types of photochemical reactions of excited metallotetrapyrroles are classified. As criteria for the classification, both the nature of the primary photochemical step and the net overall chemical change, are taken. Each of the classes is exemplified by several recent results, and discussed. The data on exciplex and excimer formation processes involving excited metallotetrapyrroles are included. Various branches of practical utilization of the photochemical and photophysical properties of tetrapyrrole complexes are shown. Motives for further development and perspectives in photochemistry of metallotetrapyrroles are evaluated.

List 1 Abbreviations for Tetrapyrroles

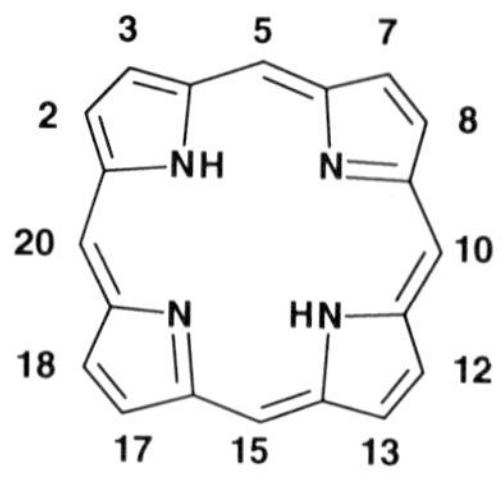

H_2 (Por)

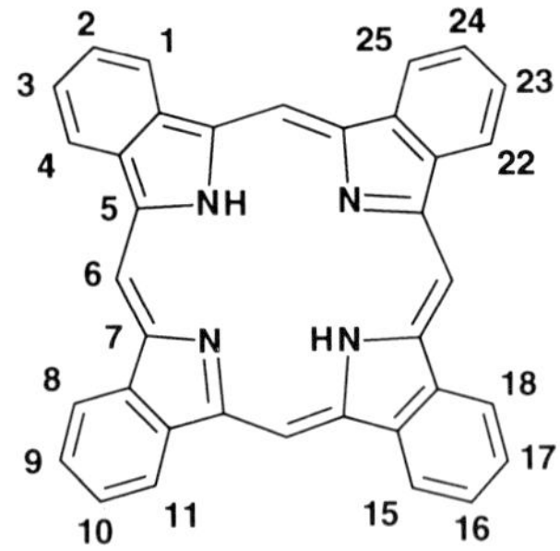

H_2 (Pc)

Substitution pattern of the ring-protonated porphyrins H_2 (Por) and phthalocyanines H_2 (Pc). In the quoted tetrasubstitued porphyrins the substituents are in 5, 10, 15, 20 meso positions, in octasubstituted porphyrins they are in 2, 3, 7, 8, 12, 13, 17, 18 positions. In tetrasubstituted phthalocyanines the substituents are in 2, 9, 16, 23 positions, in octasubstituted complexes they are in 2, 3, 9, 10, 16, 17, 23, 24 positions. Metal complexes are specified by replacing the two H atoms in H_2 (Por) or H_2 (Pc) by the metal atom with its axial ligands, if any.

H_2 (DOP)	2,12-Di(n-octyl)porphyrin
H_2 (DOEPIX)	Mesoporphyrin-IX-dioctadecyl ester
H_2 (DPDME)	Deuteroporphyrin-IX-dimethylester
H_2 (Etio I)	Etioporphyrin I
Hb	Hemoglobin
Mb	Myoglobin
H_2 (Nc)	Naphthalocyanine
H_2 (OEP)	Octaethylporphyrin
H_2 (OECP)	Octaethylporphycene
H_2 (OOOP)	Octakis(β-octyloxyethyl)porphyrin
H_2 (OSPc)	Tetrakis(octadecylsulphonamido)phthalocyanine
H_2 (Pc)	Phthalocyanine
H_2 (PcTMS)	Tetra(sulphomorpholide)phthalocyanine
H_2 (PCOO)	13,17-diethyl-3,7,8,12,18-pentamethylporphine-2-acetic acid
H_2 (Por)	Porphyrin (in general)
H_2 (PPIX)	Protoporphyrin IX
H_2 (PPIXDME)	Protoporphyrin-IX-dimethyl ester
H_2 (TAFP)	Tetrakis(aminophenyl)porphyrin
H_2 (TAP)	Tetrakis(4-methoxyphenyl)porphyrin
H_2 (TMAPP)$^{4+}$	Tetracation of tetrakis(trimethylaminophenyl)porphyrin
H_2 (TBPc)	Tetrabutylphthalocyanine
H_2 (TBPP)	Tetrakis(3,5-*tert*-butyl-4-hydroxyphenyl)porphyrin

$H_2(TClMP)^{4+}$	Tetracation of tetrakis(3,5-dichloro-1-methyl-4-pyridinio)porphyrin
$H_2(TDClP)$	Tetrakis(2,6-dichlorophenyl)porphyrin
$H_2(TdPc)$	Tetrakis(dodecylsulphonamido)phthalocyanine
$H_2(TFPP)$	Tetrakis(pentafluorophenyl)porphyrin
$H_2(TMEP)$	Hematoporphyrin-IX-dimethylether-dimethylester
$H_2(TMOP)$	2,7,12,17-Tetramethyl-3,13-dioctylporphyrin
$H_2(TMP)$	Tetrakis(mesityl)porphyrin
$H_2(TMPyP)^{4+}$	Tetracation of tetrakis(*N*-methyl-4-pyridinio)porphyrin
$H_2(TOOPP)$	Tris(octyloxyphenyl)porphyrin
$H_2(ToPc)$	Tetrakis(octylsulphonamido)phthalocyanine
$H_2(TP)$	Tetrapyrrole (unspecified)
$H_2(TpivPP)$	Tetrakis(α,α,α,α-o-pivalamidophenyl)porphyrin
$H_2(TPP)$	Tetraphenylporphyrin
$H_2(TPPC)^{4-}$	Tetraanion of tetrakis(4-carboxyphenyl)porphyrin[a]
$H_2(TPPS)^{4-}$	Tetraanion of tetrakis(4-sulphonatophenyl)porphyrin[a]
$H_2(TPyP)^{4+}$	Tetracation of tetrakis(4-pyridinio)porphyrin
$H_2(TSPc)^{4-}$	Tetraanion of tetrasulphophthalocyanine[a]
$H_2(TTP)$	Tetrakis(4-tolyl)porphyrin
$H_2(UroP)^{8-}$	Octaanion of uroporphyrin[a]

List 2 Abbreviations for Axial Ligands and Other Species

ACN	Acetonitrile
AF	Azaferrocene
BMA	Butyl methacrylate
BQ	1,4-Benzoquinone
DMF	*N*,*N*-dimethylformamide
DMSO	Dimethylsulphoxide
H_4EDTA	Ethylenediaminetetraacetic acid
Im	Imidazole
MeIm	2-Methylimidazole
Me_2Im	1,2-Dimethylimidazole
MePy	4-Methylpyridine
MeTHF	2-Methyltetrahydrofuran
MMA	Methyl methacrylate
MV^{2+}	Dication of methyl viologen
NADH	Nicotinamide adenine dinucleotide (reduced form)
pip	Piperidine

[a] All carboxylato and sulphonato groups dissociated

py	Pyridine
SDS	Sodium dodecylsulphate
THF	Tetrahydrofurane
tz	Tetrazine

List 3 Abbreviations for the Ground and Excited States

CT	Charge transfer (in general)
CTTS	Charge transfer to solvent
GS	Ground state
IL	Intraligand state
LF	Ligand field
LMCT	Ligand to metal charge transfer
MLCT	Metal to ligand charge transfer

1 Introduction

Photochemistry is the branch of chemistry that deals with the causes and courses of chemical deactivation processes of electronically excited particles, usually with the participation of ultraviolet, visible, or near-infrared radiation [1]. The photochemist is interested in both the modes of excited-state formation processes (direct photoexcitation, energy transfer, etc.) and the deactivation pathways of excited atoms, molecules, and ions.

Current photochemical research is strongly linked with the study of photophysical behavior of excited particles. Data on photophysical processes (such as luminescence, internal conversion, intersystem crossing, intramolecular energy dissipation) assist photochemists in the identification and interpretation of chemical deactivation modes. Most of the data related to the elementary steps within deactivation of excited particles have been obtained by fast flash techniques in nano-, pico-, and femtosecond time domains. Photophysics is, in general, as rich a branch of science as photochemistry, and both the parts of excited-state research deserve comparable attention and extent. In the present review, some results on photophysics will be mentioned where suitable and necessary. We will restrict our discussion, however, predominantly to photochemical behavior of metallotetrapyrroles.

To elaborate all aspects of the photochemistry of a class of compounds in detail, especially for such a wide range of compounds as tetrapyrrole complexes, is a task which can hardly be accomplished in the narrow confines of one chapter. Fortunately, a number of segments of the beautiful mosaic representing the chemistry of metallotetrapyrroles is known. Besides, two preceeding volumes of this series [2, 3], excellent monographs and reviews dealing with biological aspects of natural metallotetrapyrroles and their models, including the function and consequence of light absorption, have recently been published [4–22]. Data on the spectral and photophysical properties of the said compounds are also available [2, 4, 23–28]. On the other hand, less attention has been devoted to "inorganic aspects" of metallotetrapyrrole chemistry. The above facts allow us to focus our attention in writing this article on:

1) synthetic metallotetrapyrroles, in the nature and properties of which the central atom plays an important role;
2) photochemical deactivation pathways of excited tetrapyrrole complexes.

Information on natural complexes and their photoinduced biological functions are included where appropriate or necessary for the sake of completeness as well as for the possibility of comparing the differences and common features in the behavior of natural and artificial compounds.

Even a short glance at the chemical literature is sufficient to get an impression (and it is not only an impression but a reality) that during the two last decades, the photochemistry of tetrapyrrole complexes belongs to the fastest developing areas in the field of excited-state chemistry of coordination compounds and organometallics. As far as the motives for such development are concerned, some apparent contradictions seem to exist. Of the biologically active natural tetrapyrrole complexes only one kind, magnesium containing chlorophylls, are involved in natural photochemical processes. The majority of artificial compounds (e.g. metallophthalocyanines) are practically important just due to their photochemical stability (they are strong light absorbers and undergo deactivation predominantly by photophysical modes).

Assessing interrelations between the practical importance (as a fundamental driving force for any research) and the research activity in the field of photochemistry of metallotetrapyrroles from the other side, photosynthesis is such an extremely important natural process that it is itself a sufficient stimulus for the enormous development in photochemistry of coordination compounds. Photosynthesis is a key driving force in the study of energy transfer and dissipation within one molecule or organized systems, electron transfer processes, charge separation and its stabilization. Other driving forces stem from the stability, rigidity, and ability of metallotetrapyrroles to act in light-driven industrial (e.g. photocatalysis, photovoltaic cells) and biological processes (e.g. photodynamic therapy). It is worth mentioning, however, that the levels of knowledge obtained and the conclusions drawn depend on the type of complexes and problem investigated. Along with very profound studies, there are still vast blank areas in the field of photochemistry of metallotetrapyrroles.

Most results, particularly those obtained in the last decade, have been attained by fast flash techniques, usually using various laser sources of monochromatic radiation. Such methods can provide information on the elementary photochemical and photophysical deactivation steps. For some practical applications, it is more favorable to work with continuous and/or polychromatic radiation. In this article both kinds of results are evaluated.

The scope of this review is threefold. Firstly, to present all known types of photochemical reactions of artificial metallotetrapyrroles and to classify them according to the primary photochemical deactivation modes, and net chemical changes. Secondly, to point to the broad extent of knowledge being gained in the field and to the areas still waiting for systematic photochemical research. The latter point can be understood as a hint for further research orientation. Thirdly, to provide the reader (especially those engaged in the research areas overlapping with photochemistry of metallotetrapyrroles) with a fair background in the reviewed matter and its significant applications.

Turning back to the definition of photochemistry and anticipating the classification of photochemical reactions of metallotetrapyrroles, it should be kept in mind that a true photochemical process is only that involving an electronically excited particle (in this review it means an excited tetrapyrrole complex). All subsequent reactions are spontaneous (in photochemistry they are familiarly called "dark" reactions). Exactly speaking, each classification of photochemical reactions should start with an answer to the question: "what is the nature of the primary photochemical step involving a complex in its photochemically reactive excited state?" It must be admitted that for the majority of photochemical reactions only indirect evidence exists on both the photoreactive state and the primary photochemical step. The reason for this situation was thoroughly discussed in our previous work [1]. There is a clearer situation in the field of photophysical radiation deactivations.

Photochemists who rely more on experimental data and proofs than on hypotheses know that there are still blank areas in solving the most general problem of photochemistry: interrelationships between the nature of the photoreactive excited state of a molecule and the pathways and efficiency of its chemical deactivations [29]. The categorization of photoreactions of metallotetrapyrroles, presented in this article, stems from available published data, and the opinions and deductions of their authors.

2 Exciplex and Excimer Formation

Changes in the electronic and molecular structures of a molecule A due to a transition from its ground to an excited state can result in creation of conditions for chemical bonding between the excited molecule *A (in the whole review the symbol *A will denote an excited particle A in general; the number X in the symbol ^{X}A denotes the multiplicity of the excited particle A), and another molecule Q of the system, giving rise to an excited adduct *(A – Q) [1]. Such an adduct formed in a bimolecular dynamic adiabatic process

$$ {}^{*}A + Q \Rightarrow {}^{*}(A - Q) \tag{1} $$

is called an exciplex (if the molecules A and Q are different in the ground state) or an excimer (if the molecules A and Q are identical). Both terms are conventionally used in cases when the molecules A and Q being in the ground state do not form any adduct. The formation of exciplexes and excimers belongs to the simplest photochemical reactions, since besides a bond formation no substantial changes in the molecules A and Q occur and the molecules preserve, at least partly, their chemical identity in the adduct.

It should be mentioned that, in principle, both the molecules (*A, *Q) can be excited when forming an exciplex or excimer. Moreover, along with bicomponent exciplexes and excimers *(A – Q), evidence for the formation of three-component adducts *(Q – A – Q) has been found in some systems.

Exciplexes and excimers have their own structure and properties (e.g. multiplicity, absorption and emission spectra, lifetime, stability constant, enthalpy and entropy content, pathways of deactivation) and can be regarded thus as new chemical species.

Experimental data on the formation, thermodynamic and kinetic decay parameters, and the multiplicity of exciplexes and excimers involving tetrapyrrole complexes are summarized in Tables 1 and 2. Based on these data and on information on the spectral properties and chemical behavior, some conclusions

Table 1. Formation, thermodynamic and kinetic decay parameters (stability constant K, energy and entropy values, decay rate constant *k*, lifetime τ, quantum yield ϕ) of exciplexes *(A – Q) involving tetrapyrrole complexes *A (energy values expressed in kJ mol^{-1}; entropy in J mol^{-1} K^{-1}; k and τ in s^{-1} and s, respectively)

Complex *A, its excited state	Counter partner Q	Solvent	Properties of exciplexes *(A – Q) Notes	Ref.
Mg (TPP), ^{3}IL	BQ and its derivatives, Nitroaromatics, Chlorocompounds	Toluene	K ~ 10^5; k ~ 10^3; ΔH ~ – 10^0 to 10^1; ΔS ~ 0 to 10^2; ΔG ~ – 10^1 to 10^1	[30–33]
	Duroquinone	Ethanol	T = 243 K	[34]
Mg (Etio I), ^{1}IL, ^{3}IL	Nitroaromatics	Benzene	Formation of both singlet and triplet exciplexes *(A – Q) suggested; three-component exciplexes *(Q – A – Q) in triplet state detected	[35–37]
Mg (TPP) (py) ^{3}IL	Nitroaromatics, Chlorocompounds, Quinones	Toluene	k ~ 10^2–10^4	[38, 39]
Zn (TPP) ^{1}IL, ^{3}IL	BQ, Duroquinone, O_2, 4–X–$C_6H_4N_2^+$ (X = H, Cl, Br, CH_3, OCH_3, C_4H_9, $N(C_2H_5)_2$)	Acetone, ACN, Toluene, DMF Dioxane, CH_2Cl_2, Ethanol (243 K)	$\tau = 10^{-5}$–10^{-6}; K = 10^3–10^4; ϕ_{phosp} = 0.39 – 0.88; singlet exciplex formed when Q = $^1O_2$	[40–44, 34]
	1,4-Dinitrobenzene	Hydrocarbons	ΔH = 18	[32]
Zn (Etio I), ^{3}IL	Nitroaromatics, Chlorocompounds	Benzene	k ~ 10^4–10^5; formation of both triplet *(A – Q) and *(Q – A – Q) evidenced	[35, 36, 39, 40]
Zn (MPIXDME) ^{1}IL, ^{3}IL	Nitrostilbenes	Benzene	Formation of both singlet and triplet exciplexes *(A – Q) suggested	[35, 36]
Zn (TMPyP), ^{3}IL	Methylviologen	Water		[45]
Zn (OEP), ^{3}IL	Nitroaromatics, Chlorocompounds	Benzene		[37]
Al (TPP) (OH), ^{3}IL	Nitroaromatics, Chlorocompounds, Quinones	Toluene	K ~ 10^5; k ~ 10^3–10^5	[30, 33, 38, 39]
Ga (TPP) (OH), ^{3}IL	Nitroaromatics, Picryl chloride	Toluene	k ~ 10^3–10^4; for Q = dinitrobenzene, ΔG = 26, ΔS = 107, and ΔH = – 5.9	[30, 31, 38, 39]
In $(TPP)^+$, ^{3}IL	Methylviologen, MV	Methanol	K = 6.5 × 10^2; MV assumed to interact with porphyrin ring when forming the exciplex	[46]

In (TPP) $(MV)^{3+}$ ^3IL	Triethanolamine, TEA	Methanol	TEA occupies an axial position in the exciplex	[46]
In (TPP) (OH), ^3IL	Benzoquinones, Naphtoquinones	Toluene	$K \sim 10^5$–10^6; $k \sim 10^3$–10^4	[30, 33, 38, 39]
In (TPP) (C_2H_5) ^3IL	Pyridine	Benzene	$K = 6.45 \times 10^2$; $k = 7.5 \times 10^4$	[47]
Sn $(TPP)Cl_2$ ^3IL	9-Phenyl-10-methyl-9,10-dihydro-acridine	$C_2H_4Cl_2$/ACN 1:1 (v/v)	$K = 6.5 \times 10^2$; $\tau = 4.6 \times 10^{-5}$; $k = 2.18 \times 10^4$	[48]
Cu (TMPyP), ^2IL	Poly(dA – dT). Poly(dA – dT), DNA	Water	The A – T site of the nucleic acid participates in the exciplex formation; $\tau > 10^{-11}$	[49]
Pd (TPP), ^3IL	*N,N*-Dimethylaniline	Benzene, Pyridine, Acetone, 2-Propanol, Isobutyronitrile	Emitting non-polar exciplex with room temperature phosphorescence; $K \sim 10$; $k \sim 10^5$	[50]
Pd (Etio I) PD (OEP), ^3IL	Ni (TPP)	Toluene	$k \sim 10^6$–10^8	[51]
Mg (TBPc), ^3IL	Quinones, Nitroxyl radicals	Toluene	$K \sim 10^4$–10^5; $k \sim 10^3$	[30, 33, 38, 39, 52]
Zn (TBPc), ^3IL	BQ	Undecane, $(C_2H_5)_2O$, Hexane	$k = 1.3 \times 10^4$ in hexane	[53]
Zn (PcTSM), ^3IL	4-CH_3O-$C_6H_4N_2^+$	ACN	$\tau = 1.57 \times 10^{-4}$; dipole moment of the exciplex is (4.8 D) smaller than that of Q (9.4 D)	[54, 55]
Al (TBPc)Cl, ^3IL	Quinones	Toluene	$k \sim 10^3$	[30, 38, 39, 52]
Al (TBPc)(OH), ^3IL	BQ	Toluene	$K = 10^6$; $k = 10^4$	[30]
Ru (Pc)LL′, ^3IL (L,L′ = DMSO, DMSO; DMF, CO; py, CO; CH_3OH, Cl)	Dinitrobenzene, Paraquat, Diaquat	ACN	$k \sim 10^5$–10^6	[56]
Pc–Ln–PcH, ^3IL (Ln = La, Nd, Lu, Y)	CH_2Cl_2	ACN/CH_2Cl_2	Exciplex formation suggested	[57]
Chlorophyll *a* Chlorophyll *b* ^3IL	Nitroaromatics, Quinones, Benzaldehyde	Hydrocarbons, Alcohols, DMF, $(C_2H_5)_2O$,	$\tau \sim 10^{-4}$–10^{-6}	[58]

Table 2. Excimers formed by tetrapyrrole complexes

Excimer-forming complexes (GS = the ground state)	Multiplicity of excimer	Solvent	Formation (K) and decay rate (k, in s^{-1}) constants	Ref.
$^3Zn(Por) + {}^3Zn(Por)$ (Por = TPP or OEP)	1	Benzene	$k \sim 10^3$	[59]
$^3Al(Pc)Cl + {}^3Al(Pc)Cl$	1	Cyclohexane	$k = 8.4 \times 10^3$	[59, 60]
$^3Si(Pc)(OC_2H_5)_2 + {}^3Si(Pc)(OC_2H_5)_2$	1	1-Chloronaphtalene	$k = 3.8 \times 10^3$	[59, 60]
^3Pd (TPP) + $_{GS}^{1}Pd$ (TPP)	3	Toluene, THF, Mineral oil	$K = 10^3$ (in toluene)	[61–63]
^3M (Etio I) + $_{GS}^{1}M$ (Etio I) (M = Zr or Pd)	3	Mineral oil, THF	$K \sim 10^{-1}$; $k \sim 10^7–10^8$	[61]

concerning exciplexes and excimers in which A = tetrapyrrole complexes can be drawn.

In the majority of known cases, the triplet excited state is characteristic for both the excited complexes ^3A participating in exciplex formation, and the exciplexes $^3(A - Q)$ having been formed. This is probably a consequence of a longer lifetime and lower energy content in the lowest lying spin-forbidden triplet as compared with spin-allowed singlet excited states. It should be mentioned, however, that the phosphorescence at room temperature in solution is a rather rare event and evidence on the exciplex properties is mined from quenching kinetic data, electronic absorption spectra of excited particles (ESA spectra), and chemical reactions occurring in irradiated systems.

A further conclusion concerns the efficiency of intersystem crossing processes (intersystem crossing is an isoelectronic radiationless transition between two excited states having different multiplicities). For complexes the quantum yield of a singlet → triplet intersystem crossing approaches unity and the process is a very common one [1]. Based on the experimental observations which are, however, available for a few cases only, singlet and triplet exciplexes do not convert and the absence of the intersystem crossing in them seems to be a general phenomenon [30].

A characteristic feature of triplet exciplexes $^3(A - Q)$ is that they are quenched via energy transfer by molecules Q′ with a low-lying triplet level (e.g. dioxygen, azulene, tetracene) and the corresponding quenching rate constants k_q reach the diffusion limited value k_{dif}.

It should be pointed out that when studying literature sources one must distinguish between the bimolecular quenching of exciplexes $^3(A - Q)$ by a quencher Q′ (related k_q has usually the value of the order 10^8–10^{10} $M^{-1} s^{-1}$)

$$^3(A - Q) + Q' \xrightarrow{k_q} A + Q + {}^3Q' \tag{2}$$

and the rate constant of monomolecular exciplex decay

$$^3(A - Q) \xrightarrow{k_q} A + Q \tag{3}$$

whose value is of some order lower ($k_q \sim 10^3$–10^5 s^{-1}).

Exciplexes of metallotetrapyrroles are usually of a charge transfer nature with an excited complex *A acting as an electron donor and a molecule Q playing the role of an electron acceptor. In most instances the covalent contribution to the stability of exciplexes cannot be neglected. The extent of the electron transfer (expressed as the change in the molecular dipole moment) can be roughly estimated from the electronic absorption spectra of exciplexes (measured by nano- or picosecond transient spectrophotometry): for a given *A and solvent, the greater the electron affinity of the quencher Q, the higher energy of the absorption band maximum of the exciplex *(A – Q). The effects of the electron-donating ability of *A and of solvent polarity are complex and cannot be predicted in a simple way.

The thermodynamic stability of exciplexes, expressed through their formation constant K_f

$$*A + Q \Rightarrow *(A - Q) \qquad K_f = [*(A - Q)]/[*A]\cdot[Q] \tag{4}$$

reflects both the enthalpy and entropy contributions which are comparable in extent. At first glance the positive value of entropy contributions are not so easily understood as the exciplex formation is associated with a loss of some rotational degrees of freedom and the decrease of the number of molecules. These factors, lowering the entropy, are overcome by the entropy gain due to newly formed vibrational degrees of freedom and changes in solvation.

Generally, the differences in Gibbs energy, ΔG, of a solvated exciplex *(A – Q) and its solvated constituents *A and Q are not very large. The values of the formation constants K_f for the exciplexes involving a tetrapyrrole complex are of the order 10^3–10^5 (for triplet exciplexes) and of some order lower for singlet exciplexes.

The problem of the formation and existence of exciplexes is connected with their character (predominantly charge-transfer or covalent) and the polarity of solvent. Knowledge in this matter can be summarized as follows: the higher the polarity of the solvent (its dielectric constant), the lower is the stability of charge-transfer exciplexes in it. Charge-transfer exciplexes are stabilized, therefore, in non-polar solvents. For the stability of non-polar exciplexes (a rare case of exciplexes containing a tetrapyrrole complex) the solvent polarity plays a negligible role. As it follows from the data in Table 1, there are some exceptions to the above rules.

The problem of the decay of exciplexes arises directly from their stability. There is no simple and unambigous relation between the electron donor ability of an excited complex *A, electron withdrawing property of its reaction partner Q, and the decay rate constant k_q. This experimental observation may be understood on the basis of two deactivation pathways of exciplexes *(A – Q): a spin-forbidden conversion to the singlet ground state (GS) molecules A and Q, e.g. for a triplet exciplex $^3(A - Q)$

$$^3(A - Q) \Rightarrow {}^1_{GS}A + {}^1_{GS}Q \tag{5}$$

and a spin-allowed dissociation of the exciplex followed by a quenching of the excited complex, e.g.

$$^3(A - Q) \Rightarrow {}^3A + {}^1_{GS}Q \Rightarrow {}^1_{GS}A + {}^1_{GS}Q \tag{6}$$

Both the decay channels are realized independently. The decay in which the excited quencher *Q would be formed is usually largely endoergonic and, therefore, such a pathway of degrading the electronic energy is implausible.

Internal exciplexes and excimers (excited molecule in which one part bearing a positive charge interacts with another part bearing a negative charge, both charges being created as a consequence of a photoexcitation of the molecule) known for organic compounds such as diphenylalkanes [30] have been described neither in the photochemistry of metallotetrapyrroles in particular, nor in that of metal complexes in general. There is no reason, however, why internal exciplexes or excimers should not exist in the above-mentioned class of metal atom containing compounds (e.g. their existence can be supposed in the case of cofacial dimeric or polymeric metallotetrapyrroles).

A few data dealing with excimer formation and properties of tetrapyrrole complexes have been published. A lack of observation in this field is typical for inorganic photochemistry as a whole. Excimers are formed from excited and ground-state complexes (as an example the case of palladium(II) containing excimers can be introduced [61–63]) or from two excited complexes. The first (and to the author's knowledge the only) evidence of the excimer formation by self association of two excited tetrapyrrole complexes came from Ferraudi's laboratory in 1986 [59–60]. Based on the very detailed study including the investigation of the influence of the solvent polarity and external magnetic field on excimer formation, the authors claimed that the singlet excimers are formed by an interaction of two triplet complexes. The formation of one of the excimer, $^1[Al(Pc)Cl]_2$, can be expressed as follows

$$2\,^3[Al(Pc)Cl] \Rightarrow {}^1[Al(Pc)Cl]_2 \tag{7}$$

The bonding in the excimers is covalent in nature. In polar solvents the excimers dissociate forming redox changed radicals, the unpaired electron being localized on the phthalocyanine ring, e.g.

$$^1[Al(Pc)Cl]_2 \Rightarrow {}^2_{GS}[Al(Pc^{\bullet})Cl]^- + {}^2_{GS}[Al(Pc^{\bullet})Cl]^+ \tag{8}$$

In solvents with low polarity, the exciplexes dissociate to a monomeric singlet excited complex and a singlet ground-state complex, e.g.

$$^1[Al(Pc)Cl]_2 \Rightarrow {}^1[Al(Pc)Cl] + {}^1_{GS}[Al(Pc)Cl] \tag{9}$$

It was stated in our previous review [1] that exciplexes and excimers are not very common in the photochemistry of coordination compounds, probably due to the fact that they have not been systematically looked for. This part can be concluded with a statement that only a small change has occurred in this matter.

3 Photosubstitution, Photoelimination, and Photoaddition Reactions

This Section deals with photochemical deactivations in which the primary photochemical step is a non-redox, heterolytic central atom – ligand bond breaking, and in the final products both the central atom and the tetrapyrrole ligand retain their oxidation state.

A photosubstitution reaction can be expressed as follows

$$X - M(TP) - L + L' \xrightarrow{h\nu} X - M(TP) - L' + L \tag{10}$$

where M is the central atom; L and L′ are the leaving and entering ligand, respectively; TP is the tetrapyrrole ligand; X is the other monodentate (if any) ligand coordinated to the central atom. Usually photosubstitutions in metallotetrapyrroles consist of two chemical steps, the first one being the elimination of the leaving ligand L from the excited complex

$$X - M(TP) - L \xrightarrow{h\nu} X - M(TP) + L \tag{11}$$

followed by the addition (thermal step) of an entering ligand L′ to the coordinatively unsaturated complex X – M(TP)

$$X - M(TP) + L' \longrightarrow X - M(TP) - L' \tag{12}$$

In most cases, the two latter processes have been studied individually by fast techniques (flash photolysis, transient spectra measurements, Raman spectroscopy) in nano-, pico-, and femtosecond time scales as processes accompanying photophysical deactivation steps [64–66]. In Table 3 the data for such individual steps are reported. The data can be summarized as follows:

1) photoexcitation results in an ejection of one ligand from the primary coordination sphere and in noncoordinating solvents (e.g. CH_2Cl_2, hydrocarbons), either species with a lower coordination number are formed, or a very rapid recoordination of the ejected ligand occurs;
2) in coordinating solvents (e.g. pyridine, piperidine, DMSO) a solvent molecule is trapped by the unsaturated complex and, from the viewpoint of stoichiometry, a typical substitution reaction proceeds;
3) in the majority of the systems studied, L = L′ and the systems can be characterized as photochromic (photochromism means a photoinduced transformation of a molecular structure, photochemically or thermally reversible, that produces a spectral change, typically in the visible region).

In addition to photosubstitution and photoelimination reactions, in the cases of some Ni(II) complexes, photoexcitation of square-planar complexes Ni(TP) and formation of the photoassociative ligand-field (LF) excited state $^3B_{1g}$ can lead to photoaddition reactions yielding hexacoordinate complexes $Ni(TP)L_2$ [65, 66, 75–77]. Such processes differ from the second step of photosubstitutions since an excited complex participates in them and the addition is conditioned by the electronic structure of the complex in its excited state (see Table 3).

It may be worth mentioning that several reactions are known, which, from the viewpoint of stoichiometry of the overall chemical change, are photosubstitutions, e.g. [97]

$$(TPP)Al - CH_3 + HOR \xrightarrow{h\nu} (TPP)Al - OR + CH_4 \tag{13}$$

where the symbol HOR represents 2-*tert*-butyl-4-methoxyphenol. The primary photochemical step of such reactions is, however, the homolytic splitting of a central atom–ligand bond. Ostensible non-redox photosubstitution is thus a consequence of consecutive redox reactions. This kind of processes is discussed in Sects. 6 and 7.

In the photochemistry of coordination compounds and organometallics containing mono or bidentate ligands, photosubstitutions occupy virtually the first position in research activity and a number of published papers [1, 98, 99]. Photosubstitutions are those processes for which the first rules (Adamson rules [100] enabling photochemists to predict the course and relative efficiency of photoreactions) of photoreactivity in the field of inorganic photochemistry were formulated. Yet there are still a lot of questions to be answered.

On the other hand, photosubstitutions of tetrapyrrole complexes are rare processes. The main reason for such distinctions between tetrapyrrole complexes and common inorganic compounds lies in the electronic structure of the two mentioned classes of compounds in their low-lying photoreactive excited states.

For most inorganic transition metal d^1-d^9 complexes the lowest excited state is a metal-centered ligand field (LF) state of dissociative character possessing a populated antibonding d-orbital (e.g. $3d_{z^2}$ orbital in octahedral cobalt(III) complexes). In such a state some of the central metal–ligand bonds are weakened and easily undergo a heterolytic splitting. This splitting is usually the first step in photosubstitutions, non-redox photoeliminations, and photoisomerizations [1].

For phthalocyanines, regardless of transition or non-transition metal nature of the central atom, the lowest lying excited state is usually an intraligand (π, π^*) triplet [4]. The tendency of axial ligands to be released from the primary coordination sphere in this state does not substantially differ from that in the ground state. The other tetrapyrrole complexes (porphyrins, corrins, chlorins, etc.) of redox stable central ions, such as Zn(II), Mg(II), Al(III), have the same nature of the lowest lying excited state. All complexes belonging to this category undergo predominantly tetrapyrrole ring – localized photoredox processes when being deactivated chemically.

Transition metal tetrapyrrole complexes (except phthalocyanines) exhibit photochemistry similar to both typical organic compounds (e.g. they form exciplexes and excimers with a ring–ring interaction) and inorganic complexes (e.g. they provide metal localized redox and substitution photoreactions). Often the energy order of excited states depends on the nature of the central atom, axial ligands, and peripheral ring substituents. This order can be changed on purpose by the changes in the mentioned factors or complex environment (e.g. solvent polarity).

Almost in all cases (Table 3) photosubstitution reactive excited states are longer-lived spin-forbidden states involving an antibonding d-orbital of the central atom (usually d_{z^2}). The ability of complexes to undergo a photoejection of one axial ligand stems from the dissociative character of such a state. A special attention should be paid to complexes in metal-to-ligand charge-transfer (MLCT) reactive excited states (d, π^*). On one side, the increase of positive

Table 3. Photosubstitution, photoelimination, and photoaddition non-redox reactions of tetrapyrrole complexes

Complex and its photoreactive excited state	Ligand leaving	Ligand entering	Conditions and parameters of reactions Notes	Ref.
Cr(TPP)Cl(py), IL ($^4S_1$)	py	py	Solvent: acetone; $\lambda_{irr} = 355$ or 532 nm; $\phi = 0.16$, independent on λ_{irr}; $\bar{k} = 4.2 \times 10^2\ M^{-1}\ s^{-1}$	[67]
Cr(TPP)Cl(acetone), IL ($^4S_1$)	acetone	acetone	Solvent: CH_2Cl_2, acetone; $\phi = 0.63$ at $\lambda_{irr} = 355$ nm, and 0.55 at $\lambda_{irr} = 532$ nm;	[68]
M(Por)(NO), LF or MLCT M = Fe, Co, Mn; Por = TPP, OEP	NO	NO	Solvent: benzene; $\lambda_{irr} = 355$ or 532 nm; for M = Co: $\phi = 1.0$ for both Por and λ_{irr}; for the other complexes $\phi = 0.50$–0.78 and depend on λ_{irr}	[69]
Fe(Por)L_2, $^3LF(d_\pi, d_{z^2})$ Por = TPP, PPIXDME, DPDME L = imidazoles or pip	pip imidazoles	pip	Solvent: toluene/piperidine; $\lambda_{irr} = 355$ or 532 nm; $\phi = 0.03$–0.08; $k \sim 10^8$–$10^9\ M^{-1}\ s^{-1}$; five-coordinate transient Fe(Por)L formed	[70]
Fe(Por)L_x, $^3LF(d_\pi, d_{z^2})$ Por = TPP – Im, x = 1; TPP, x = 2; L = CO, imidazoles, py, pip	CO imidazoles py pip	CO imidazoles py pip	Solvent: toluene; $\lambda_{irr} = 530$ nm; $\phi \sim 10^{-1}$–10^{-2}; $\bar{k} \sim 10^8\ M^{-1}\ s^{-1}$ and decreases with increasing pK(L)	[71, 72]
Co^{III}(OEP)(CN) $^3LF(d_\pi, d_{z^2})$ Co^{III}(OEP)(CN)(pic) $Co^{III}(OEP)(DMSO)^+$ $Co^{II}(OEP)(pip)_2$, $^2LMCT(\pi, d_{z^2})$	CN^- CN^- DMSO pip	CN^- CN^- DMSO pip	Solvent: toluene or CH_2Cl_2 with L; $\lambda_{irr} = 355$ or 532 nm; $\phi = 10^{-1}$–10^{-2}; $\bar{k} \sim 10^9\ M^{-1}\ s^{-1}$; five-coordinate and four-coordinate species formed	[73]
Co^{III}(TPP)X(py), $^3LF(d_\pi, d_{z^2})$	py	py	Solvent: benzene/pyridine; $\lambda_{irr} = 355$ or 532 nm; $\phi = 0.12$ at 355 nm, and 0.045 at 532 nm, independently on X^- and on temperature; $\bar{k} \sim 10^9\ M^{-1}\ s^{-1}$	[74]
Ni(Por), $^3LF(^3B_{1g})$ Por = PPIX, TPP, OEP, Mb, Hb, TPyP	–	Histidine 2 py 2 pip 2 Pyrrolidine	Solvent: pyridine, piperidine, pyrrolidine; $\lambda_{irr} = 406.7$ nm, 413.1 nm, or 415.1 nm; five-coordinate intermediates involved; ϕ non reported; $^3B_{1g}$ = associative state for planar complexes	[65, 66, 75–77]

Ni(Por)L_2, $^1LF(^1A_{1g})$ Por = PPIX, TPP, OEP, Mb, Hb, TPyP	2 py 2 pip 2 pyrrolidine	–	Solvent: pyridine, piperidine, pyrrolidine; $\lambda_{irr} = 406.7$ nm, 413.1 nm, or 415.1 nm; ϕ non reported; $^1A_{1g}$ = dissociative state for hexacoordinate complexes	[65, 66, 75–77]
Ru(Por)(CO)(C_2H_5OH), ^3LF Por = TPP, OEP	CO, C_2H_5OH	2 Aniline 2 DMF 2 nitrile	Solvent: Aniline, DMF, various nitriles; medium-pressure Hg-lamp; photochemical synthesis of Ru(Por)L_2	[78, 79]
Ru(TPP)(CO)(pip) ^{3}MLCT or ^{3}LF	CO	pip	Solvent: CH_2Cl_2/piperidine; for $\lambda_{irr} = 530$ nm: $\phi = 2.5 \times 10^{-6}$ at 25 °C, and 1.5×10^{-4} at 80 °C; for $\lambda_{irr} = 412$ nm; $\phi = 5.9 \times 10^{-5}$ at 25 °C, and 1.6×10^{-4} at 80 °C	[80]
Ru(OEP)(CO) ^{3}MLCT or ^{3}LF	CO	not given	Solvent: ACN; $\phi = 0.01$ at $\lambda_{irr} = 530$ nm and 0.08 at $\lambda_{irr} = 260$ nm	[81–83]
Ru(DOEPIX)(py)L L = CO, N_2, O_2	L	CO	Monolayer dry solid assemblies of the starting complexes; pyrex-filtered light; Ru(DOEPIX)(py) stable in vacuo; in the presence of L′(CO, N_2, O_2, py), Ru(DOEPIX)(py) forms Ru(DOEPIX)(py)L′	[82, 83]
Ru(OEP)(CO)L, ^{3}MLCT(d, π^*) L = py, pip, DMSO	CO	L	Solvent: piperidine, pyridine, DMSO, CH_2Cl_2, bromobenzene; $\lambda_{irr} = 366$ or 532 nm; ϕ = low, the value not given; Ru(OEP)(CO) does not eject CO in noncoordinating solvents (CH_2Cl_2, C_6H_5Br)	[84–90]
Rh(TPP)(CO)Cl, ^3LF (d_π, d_{z^2})	CO	CO	Solvent: benzene; $\lambda_{irr} = 355$ or 532 nm	[91]
Os(TPP)(CO)L, ^3IL (π, π^*) L = CH_3OH, py	CO CH_3OH	py	Solvent: pyridine; $\phi < 3 \times 10^{-7}$ at $\lambda_{irr} > 390$ nm	[92, 93]
Fe(Pc)(MeIm)(BzNC), IL(π, π^*)	BzNC BzNC	BzNC MeIm	Solvent: ACN; $\lambda_{irr} = 550$ or 600 nm; at high c(MeIm) the complex Fe(Pc)$(MeIm)_2$ is formed with $\phi = 0.038 \pm 0.001$	[94]
Rh(Pc)X(CH_3OH), ^3IL(n, π^*) X = Cl, Br, I	CH_3OH, X	2 CH_3CN	Solvent: ACN; $\lambda_{irr} = 254$ nm; $\phi \sim 10^{-2}$; no X_2^- detected; intermediates Rh(Pc – H)(CH_3OH)X formed by hydrogen atom abstraction from CH_3CN	[95]
Rh(corrin)Cl_2, $^3LF(d_\pi, d_{z^2})$ Cyanocobalamin, $^3LF(d_\pi, d_{z^2})$	Cl^- CN^-	H_2O H_2O	Solvent: water; $\lambda_{irr} = 313$ nm, 350 nm or 560 nm; for cyanocobalamin $\phi \sim 10^{-4}$ and independent on λ_{irr}; for Rh(corrin)Cl_2 $\phi = 10^{-2}$ at 313 nm	[96]

charge localized on the central atom should strengthen the central atom-axial ligand σ-bond and, in consequence, such a state is not dissociative toward σ-bonding ligands (e.g. piperidine, halogenides). On the other side, depopulation of a d_π-orbital leads to the weakening of π-back bonding and, therefore, the MLCT states should be of dissociative character toward strong π-acceptors (e.g. CO, NO^+). These states, when accessible, can thus be photoreactive states in substitutions of π-acceptor ligands. It should be mentioned that there are difficulties in identifying photoreactive excited states since the absorption spectra of tetrapyrrole complexes are dominated by very intense intraligand absorption bands. Anyway, the photochemically reactive excited states can be reached either by a direct photoexcitation and energy transfer processes, or via the chain of excited states located nearby (e.g. [96]) the photosubstitutionally reactive ligand-field triplet state $^3LF(d_\pi, d_{z^2})$ is the final state in the sequence starting with the directly populated tetrapyrrole-ring localized intraligand state, $IL(\pi, \pi^*) \rightarrow LMCT(\pi, d_{z^2}) \rightarrow LF(d_\pi, d_{z^2})$, or it can be obtained by thermal activation from a lower energy state [80].

In most papers devoted to photosubstitution reactions [104–109] the authors connect their research with biologically significant processes of binding and evolving simple diatomic molecules (O_2, NO, CO, CN^-) from hemoproteins (hemoglobin Hb, myoglobin Mb, cytochrome oxidase aa_3, peroxidases HRP, etc.). It has been long recognized that hemoproteins readily react with simple molecules and their adducts undergo photodissociation. Many partial problems have been solved investigating systematically the synthetic model analogs of hemoproteins. Instead of dioxygen adducts, CO adducts are frequently used in the research due to their redox unreactiveness and high quantum yield ($\phi \sim 1$) of CO photoejection [107]. Though this article is devoted predominantly to synthetic tetrapyrrole complexes, it seems to be useful, for the sake of completeness, to summarize the results on photochemistry of natural systems too.

Photodissociation of O_2 from HbO_2 singlet ground state adducts can be schematized as follows [103]

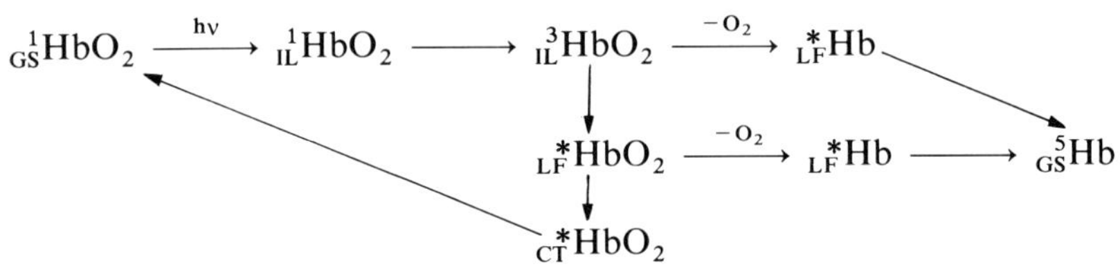

(14)

where GS are the ground states; IL are the porphyrin-ring localized intraligand (π, π^*) excited states; LF denotes the ligand field excited states; CT is the ligand-to-metal charge-transfer excited state; multiplicity is specified where known). The high-spin deligated form of Hb is formed from the triplet intraligand excited state of the adduct, $_{IL}^{3}HbO_2$, within 350 fs while the central atom adopts an out-of-plane position. The rate constant of O_2 recoordination is of the order of 10^7–10^8 $M^{-1} s^{-1}$ [105]. In the case of CO-adducts the related rate constant is higher, reaching 10^8–10^9 $M^{-1}s^{-1}$. The removal of released O_2 or CO in the natural system is strongly limited by a protein pocket.

There is no general consensus on why the difference in the quantum yield of photosubstitutions is so large for O_2-adducts ($\phi \sim 10^{-3}$) and CO-adducts and on which excited states are responsible for this difference. An explanation based on a different efficiency of the recoordination of released O_2 or CO molecules (geminate recombination) can be ruled out, as in the systems with the same biocomplex (e.g. HbO_2 and HbCO) both molecules (O_2 and CO) have nearly identical escaping probability from the protein cage due to their similar size, mass and polarity. The reason could, therefore, lie in the different photoreactive excited states involved.

Low yields of dioxygen elimination are probably associated with very fast relaxation of nondissociative charge-transfer states (time scale of less than 15 ps) by radiationless pathways. These states involve a promotion of Fe 3*d* electrons between MO's composed mainly of Fe 3*d* orbitals and $O_2(\pi_g)$ orbitals. For CO-adducts, the lowest excited states appears to be a ligand-field triplet ($^3T_1$) or quintet ($^5T_2$) with dissociative nature owing to the promotion of an electron to the Fe $3d_{z^2}$ orbital. Such a reactive state lies below the porphyrin localized triplet $^3(\pi, \pi^*)$. Thus a competition of rapid radiationless decay occurring from low-lying charge-transfer states and dissociation from ligand-field states can be the main reason of the different behavior of CO- and O_2-adducts.

The application of high pressure kinetic techniques casts light on the problem from another angle. The activation volume for MbO_2 formation has a positive value ($\Delta V^{\#} = 5.2\ cm^3\,mol^{-1}$) whereas for MbCO formation it has a negative value ($\Delta V^{\#} = -10.0\ cm^3\,mol^{-1}$) [110]. It follows from these results that the rate determining step for the binding of O_2 to myoglobin is its entry into the protein whereas for binding of CO it is the iron–carbon bond formation.

Irradiation of a CO-adduct of cytochrome oxidase induces the transfer of a CO molecule, originally bound to an iron(II) center, to copper(I) within 1 ps [108]. Photoinitiated dissociation of CO actually occurs in less than 100 fs, probably on the time scale comparable with a vibrational period of the Fe–CO stretch (~ 64 fs). Since the involved reaction centers, Fe(II) and Cu(I), are very close, the mechanism in which the Cu–CO bond begins to form as the Fe–CO bond breaks, seems to be a plausible description of the CO transfer. A similar mechanism is believed to hold for the O_2 transfer in biological processes in which cytochrome oxidase participates.

In many natural hemoproteins the central iron atom is coordinated by an imidazole nitrogen atom of a histidine residue, connected by a protein chain with the heme. Dynamics of the elementary steps involving histidine (Hist) is expressed for CO-adducts as follows [105]

$$\begin{array}{ccc} \mathrm{Fe(Por)(Hist)(CO)} & \xrightarrow[300\ \mathrm{fs}]{h\nu} & \mathrm{Fe(Por) + Hist + CO} \\ 10^{-2}\ \mathrm{s}\ \Big\uparrow\ +\mathrm{Hist} & & 10^{-6}\ \mathrm{s}\ \Big\downarrow\ +\mathrm{H_2O} \\ \mathrm{Fe(Por)(H_2O)(CO)} & \xleftarrow[10^{-4}\ \mathrm{s}]{+\mathrm{CO}} & \mathrm{Fe(Por)(H_2O)} \end{array} \tag{15}$$

In the majority of investigated systems, steady-state irradiation leads to no net chemical change. Flash photolysis using lasers and other ultrafast techniques are needed to follow the course and dynamics of photosubstitution reactions, i.e. the elimination of an axial ligand (photosubstitutions of tetrapyrrole ligands have never been observed and, in addition, no information is available on photoinduced breaking of a tetrapyrrole donor nitrogen atom – central atom bond). Recoordination of the ejected ligand does not usually occur under diffusion control but with rates of more than one order of magnitude slower. It follows from the fact that the initial products of photoelimination (coordinatively unsaturated complex and photoejected ligand) are separated by some shells of solvent molecules before their new contact. The entropy factors associated with changes in molecular orientation within such recoordination processes have not been studied in detail so far. The same conclusion concerns the relationships between the rate constant of the recoordination of the ejected ligand and the labile or inert nature of the central atom involved (from this point of view the low values of the rate constants for recoordination in some chromium(III) complexes [67, 68] may be interesting).

The final remark on photosubstitution reactions of synthetic tetrapyrrole complexes deals with the different conclusions of various authors investigating the same or very similar systems. In some instances these differences are only apparent and they can be a consequence of different formulae writing. They may reflect, however, differences in experimental techniques used and reaction products formed (some methods provide information on intermediates, others on products stable in specified and strictly kept conditions, etc.). The complexes Ru(Por)(CO)L can serve as an example. According to paper [84] photoexcitation of the complex (expressed as pentacoordinated Rw(OEP)(CO)) in noncoordinating solvents (e.g. CH_2Cl_2, C_6H_5Br) does not lead to any CO ejection, and a five coordinated intermediate of charge-transfer nature, *Ru(OEP)(CO) is formed; in acetonitrile the complex Ru(OEP) is supposed to be formed [81], the authors did not consider a coordination of an acetonitrile molecule to the central atom Ru(II) and, in agreement with its formula, the product should be, therefore, a tetradentate square-planar complex; on the other hand, in Ref. [78] the formation of $Ru(OEP)(CH_3CN)_2$ is claimed and acetonitrile is understood to be a coordinating solvent. In addition to the mentioned reactions associated with the elimination of a CO molecule in the primary photochemical step, the paper [85] reported that the elimination of a CO molecule from Ru(OEP)(CO)(py) gave rise to the metal–metal double bonded diamagnetic dimer (py)(OEP)Ru = Ru(OEP)(py) both in pyridine (coordinating solvent) and benzene (noncoordinating solvent); reinvestigation of the problem [89], using the complex $Ru(OEP)(py)_2$ as a reactant (the complex $Ru(OEP)(py)_2$ is formed by photochemical synthesis) showed that the dimeric complex $[Ru(OEP)]_2$, formed by pyrolysis of the reactant, is a paramagnetic compound which can be converted in the presence of trace amounts of water and oxygen to the stable diamagnetic Ru(IV) μ-oxo-bridged binuclear complex [(OH)(OEP)Ru-O-Ru(OEP)(OH)]. It cannot be excluded that the complex

$[Ru(OEP)(py)]_2$ found in [85] is really formed, but due to the thermolysis of $Ru(OEP)(py)_2$ in a mass spectrometer.

A very unusual reaction, named incorrectly as photoisomerization [111] is the reaction of α-(2-oxo-1,3-dioxolan-4-yl)cobalamin which can be in a simplified mode expressed as follows

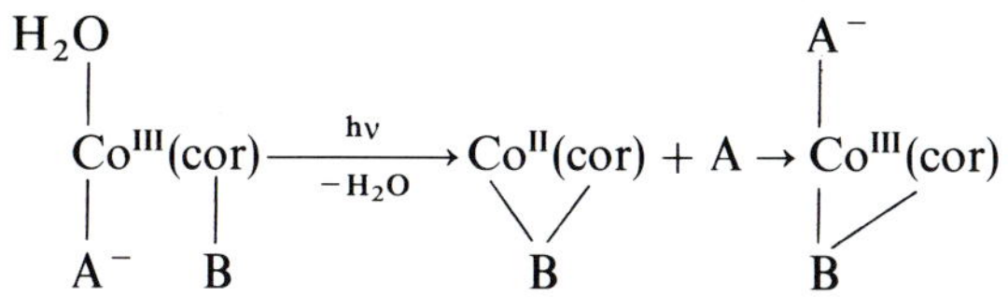

(16)

where Co(cor)-B is the cobalamin with a pending benzimidazole end-group B, A is an alkyl. From the viewpoint of stoichiometry the reaction is a photoelimination associated with a spatial rearrangement, i.e. a ring closure via coordination of the benzimidazole end-group. In reality it is a sequence of redox steps starting with a homolytic cobalt–carbon bond splitting.

Based on given experimental results and their interpretation the following conclusions on photosubstitution reactions of tetrapyrrole complexes can be drawn:

1) there are large blank areas in the photosubstitution chemistry of tetrapyrrole complexes (so far, complexes of about 10 metals have been investigated) and, vice versa, there is a wide area of research;

2) the research in this field stems from either the attempts to cast light on the mechanisms of biological processes, or to understand deactivations and energy dissipations as fundamental processes in chemistry. Photochemists are at present at the stage of collecting data. On the horizon a period appears during which the accumulated knowledge will be purposefully exploited, the course of photosubstitutions predicted and modified on purpose;

3) there is still a principal difference in the quality and unambiguousness of experimental observations and their interpretation concerning long-lived isolatable compounds and short-lived photochemically produced transients. The results on the composition, structure, and electronic state of reaction intermediates, transients, and excited states obtained by ultrafast techniques should be evaluated very carefully;

4) going from the first to the third-row transition metal (e.g. Fe → Ru → Os) a considerable drop in the quantum yields of an axial ligand photoejection and photosubstitution is seen. This experimentally demonstrated fact can be related to the ligand-field (LF) splitting and energy ordering of intraligand tetrapyrrole localized (IL) and the central atom localized (LF) excited spin-forbidden states. In the case of the first-row central atoms the lowest excited state, easily reached from directly populated energy richer IL states, is a LF state having dissociative character. For second-row metals, the energies of the lowest LF and IL states are very close and photodeactivation by physical modes (e.g. phosphorescence) becomes more effective. For third-row metals, the energy of ^3IL is substantially

lower than that of LF states (the LF splitting is the largest one) and, in consequence, such complexes are deactivated almost exclusively by phosphorescence or radiationless transition to the ground state (due to a singlet–triplet mixing by the spin–orbit coupling the triplet ^{3}IL lifetime is very short being of the order 10^{-8} s). This conclusion applies not only to tetrapyrrole complexes but to all coordination and organometallic compounds.

4 Photoinsertion Reactions

The term photoinsertion reaction applies in coordination and organometallic chemistry to the radiation-driven process in which a molecule enters the bond between the central atom and one of the ligands whereas the other central atom–ligand bonds remain unchanged. For complexes possessing a tetrapyrrole ligand (TP) the stoichiometry of photoinsertion reactions can be expressed as follows

$$X - M(TP) - L + Y \xrightarrow{h\nu} X - M(TP) - Y - L \quad (17)$$

The bond order of at least one bond in the entering molecule Y is reduced upon its insertion into the M – L bond.

Attempts to understand the mechanism of CO_2 fixation and reduction in photosynthesis gave rise to the study of photoinsertions of CO_2 by simple model compounds such as $Al(TPP)(C_2H_5)$ [112] and $In(Por)(CH_3)$ [113], where Por = TPP or OEP. For these complexes it was found that both visible light absorption and coordination of a nitrogen donor base (e.g. 2-methyl imidazole, MeIm, or pyridine, py) in an axial position are required for the insertion of CO_2 into the metal-carbon bond.

$$(MeIm)Al(TPP)(C_2H_5) + CO_2 \xrightarrow[MeIm]{h\nu} (MeIm)Al(TPP)(OCOC_2H_5) \quad (18)$$

$$(py)In(Por)(CH_3) + CO_2 \xrightarrow[pyridine]{h\nu} (py)In(Por)(OCOCH_3) \quad (19)$$

Comparing the reactants and the products, the reactions are apparently non-redox processes. Using a spin-trapping EPR technique it was shown [114] that irradiation of the complexes leads to an alkyl radical formation ($CH_3^{\bullet}$ or $C_2H_5^{\bullet}$). The efficiency of the homolytic metal–carbon bond splitting depends on the electronic properties of the other axial ligand. The ostensibly non-redox photoinsertions are thus a product of two redox reactions. As far as the photoreactive excited state is concerned, the metal–carbon bond is either indirectly activated by a $\pi \rightarrow \pi^*$ excitation localized on the tetrapyrrole ring [112] or there is an energy transfer from intraligand (π, π^*) excited states to a state involving the antibonding σ^* orbital of the metal–carbon bond [114].

A photoinduced carbon–carbon bond formation proceeds within visible light induced polymerization of methyl methacrylate in the presence of $Al(TPP)(CH_3)$. The polymerization [115] can be expressed as follows

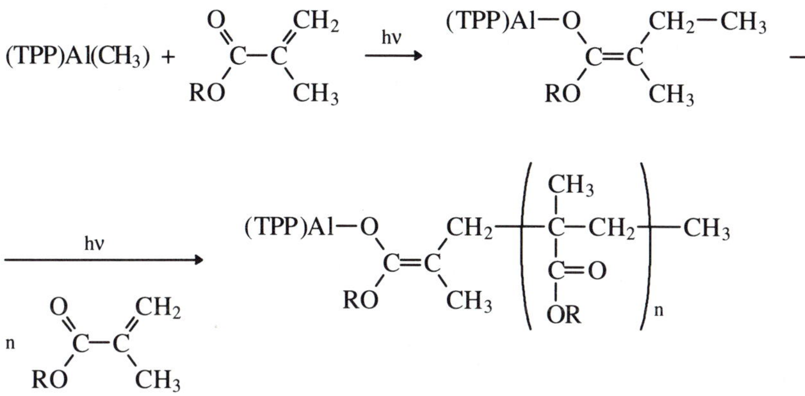

(20)

In a similar way the complex $(TPP)Al{-}O{-}C(C_4H_9) = CH{-}CH_2{-}C_2H_5$ was prepared [116] from $Al(TPP)(C_2H_5)$ and *tert*-butyl vinyl ketene together with poly-*tert*-butyl methacrylate. Every step of propagation in the photoinduced polymerization is conditioned by light absorption and includes the photo-insertion of an unsaturated molecule [115].

A metal–carbon bond splitting is the first step in the sequence leading from methylcobalamin Im–Co(corrin)–CH_3 to acetylcobalamin Im–Co(corrin)–$COCH_3$ [117]. The radical $CH_3^{\bullet}$ formed in the primary photoredox step, associated with the reduction of Co(III) to Co(II), is trapped by a CO molecule and the redox addition of the radical $CH_3CO^{\bullet}$ to the reduced pentacoordinated complex Co(II) results in the final Co(III) acetyl complex.

In connection with the bacterial synthesis of CH_3COOH and some other enzymatic insertion processes it can be envisaged that photoinsertion reactions deserve more attention in the future.

5 Photoisomerization Reactions

According to a generally accepted definition [118] isomers are individual chemical species with an identical molecular formula which display some differing physico-chemical properties and which are stable for periods of time that are long enough in comparison with those during which measurements of their properties are made. All excited states of a compound can thus be regarded as electronic and/or spin isomers, as well as distortion isomers [119] insomuch as they differ, at least slightly, in equilibrium geometry.

In chemistry, however, the term isomers is generally used in connection with isolable compounds. Keeping this limitation in mind it has been found that, apart from chemistry and photochemistry of common inorganic complexes [1], isomerism is a very rare phenomenon in the chemistry of tetrapyrrole complexes.

To our knowledge the only photoisomerization reaction leading to stable products has been reported for tetrapyrrole complexes. This reaction is [120] a ring-localized photoatropisomerization of Zn(II) and Pd(II) "picket-fence" porphyrins, (atropisomers are geometrical isomers differing in the spatial orientation of the remote groups, stable by virtue of restricted rotation about a formal single bond) occurring with low quantum yields from the lowest intraligand triplet excited state. The Cu(II) complexes do not undergo photoatropisomerization evidently due to the very short lifetime of their excited states. The formation of an isomer as a transient in laser photolysis of the complex Rh(TPP) (CO) Cl has been mentioned [91], no details were, however, given.

No data have been published on photochemical interconversion of metallotetrapyrrole optical isomers, linkage isomers, spin isomers (though such isomers are known, e.g. [Fe(TPP) (NCS) (py)] and [Fe(OEP) (3-Clpy)$_2$] ClO_4 [121]). Due to the stereochemical rigidity of tetrapyrrole rings, the types of isomer such as *cis-trans* or *mer-fac* are not expected. Sets of isomers could be synthetized having the same peripheral groups in different ring positions (positional isomers) but their mutual photochemical interconversion can hardly be expected.

6 Central Atom Localized Photoredox Reactions

Photoredox reactions of monomolecular metallotetrapyrroles can be grouped into three classes:

1) reactions involving a redox change localized on the central atom;
2) tetrapyrrole ligand localized reactions not involving the central atom (e.g. π-cation and π-anion photoformations);
3) reactions localized on the remote parts of ligands, involving neither the central atom nor the tetrapyrrole ring (e.g. redox processes localized on the protein chains of natural biocomplexes).

This section will deal with the first class of photoredox processes, two latter groups being the subject of Sect. 7.

So far, known photoredox reactions involving the central atom can be classified as follows:

a) the reactions in which the central atom localized photoredox change is associated with a redox change, addition or elimination of an axial ligand;

b) the reactions in which an axial ligand is decomposed and part of it remains coordinated to the redox changed central atom;
c) the photoinduced electron transfer between the central atom and the coordinated tetrapyrrole ring;
d) long-range electron transfer between two distant reaction sites connected by a bridging group (spacer), one of them being the central atom of a metallotetrapyrrole;
e) outer-sphere bimolecular electron transfer between an excited complex and its redox partner.

Examples of photoredox processes belonging to each of the above groups are given in Table 4. The first three types of photoredox processes (a, b, c) are mostly intracomplex monomolecular reactions (photoredox additions are bimolecular processes) which usually do not occur when a complex is in its ground state. Photoinduced long-range electron transfer reactions can be understood as a boundary between outer-sphere processes (due to a great distance between the reaction sites) and inner-sphere ones (they are, in fact, realized in one molecule).

A characteristic feature of nearly all complexes discussed in this section is the presence of a transition metal as the central atom. In contrast, tetrapyrrole ring localized redox reactions are typical for non-transition metal complexes having redox stable central atoms (e.g. Mg(II), Zn(II), Al(III)).

A light-induced redox change of the central atom and an axial ligand, followed by escaping of the redox changed ligand, can be expressed as follows

$$M^{n+}(TP)(L_{ax}^{q-}) \xrightarrow{h\nu} M^{(n-1)+}(TP) + L_{ax}^{1-q} \tag{21}$$

where M^{n+} is the central atom, (TP) denotes the tetrapyrrole ligand, L_{ax}^{q-} is the axial ligand undergoing the redox change.

Photochemical redox addition of a ligand L into an axial position can be schematized as follows

$$M^{n+}(TP) + L \xrightarrow{h\nu} M^{(n+1)+}(TP)(L^-) \tag{22}$$

The term photoredox axial ligand addition is used in instances when a stable molecule (e.g. O_2, NO) is coordinated to an electronically excited complex. As one of the rare examples of this kind of reactions, the visible-light-induced reversible formation of dioxygen adducts of Co(TPP) and its derivatives occurring in aqueous micellar solutions at room temperature, can be shown (the encapsulation of the complexes into Triton X micelles is a key factor of the reactions)

$$Co^{II}(TPP) + O_2 \xrightarrow[\text{Triton}]{h\nu} Co^{III}(TPP)(O_2^-) \tag{23}$$

The reverse process is called photoredox elimination. Superoxo cobalt(III)

Table 4. Photoredox reactions of tetrapyrrole complexes involving central atoms

Complex, its photoreactive excited state	Products	Conditions and parameters of reactions Notes	Ref.
Alkyl-Co(III) cobalamins	Co(II) cobalamins + alkylradicals	Solvent: water; the products recombine at nearly diffusion controlled rate	[122]
Methyl-Co(III) cobalamin	Co(II) cobalamin	Solvent: ethylene glycol; Co(III)–C is the strongest Co–C bond ($\Delta H = 155\ kJ\ mol^{-1}$)	[123]
$Mn^{III}(TPP)X$ $Mn^{IV}(TPP)(OCH_3)_2$ $X = Cl^-, Br^-, I^-, NCS^-, CH_3COO^-$	$Mn^{II}(TPP)$	Solvent: MeTHF, or toluene rigid matrices at 77 K; $\lambda_{irr} > 420$ nm;	[124]
$[Mn^{III}(TPP)]_2SO_4$ $Mn^{III}(TPP)(HSO_4)$	$Mn^{II}(TPP)$	Solvent: CH_2Cl_2, $CHCl_3$, hydrocarbons; $\lambda_{irr} = 350–420$ nm; $\phi \sim 10^{-4}$	[125]
M(TPP)X $M = Fe^{III}, Co^{III}, Mn^{III}, Mo^{V}O$; $X = Cl^-, Br^-, I^-, N_3^-, NCS^-$	M(TPP) + X	Solvent: MeTHF; t = 25 °C; $\lambda_{irr} > 418$ nm; $\phi \sim 10^{-4}–10^{-7}$ for Mn^{III} complexes, 8×10^{-6} for Fe(TPP)Cl, 6.7×10^{-3} for Co(TPP)Cl, $10^{-3}–10^{-5}$ for Mo^{V} complexes	[126, 127]
$Mn^{III}(Pc)(py)X$ $X = Cl^-, CH_3COO^-, OH^-$ $Mn^{III}(Pc)(CN)_2^-$ $Mn^{IV}O(Pc)(py)$ $Mn^{III}(Etio\ I)X$ $Fe^{III}(Etio\ I)Cl$	$Mn^{II}(Pc)(py)_2$ $Mn^{II}(Pc)(CN)(C_2H_5OH)^-$ $Mn^{II}(Pc)(py)_2$ $Mn^{II}(Etio\ I)$ $Fe^{II}(Etio\ I)$	Solvent: pyridine or ethanol; λ_{irr} = visible light; Mn^{II} complexes formed in the absence of O_2 (in the presence of O_2, complexes of Mn^{IV} are formed); the presence of pyridine is essential for photoreduction, the reactions are axial-ligand aided processes	[128]
$Mn^{III}(TMPyP)^{5+}$	$Mn^{II}(TMPyP)^{4+}$ $Mn^{IV}(TMPyP)^{6+}$	Solvent: water with adjusted pH; λ_{irr} = visible light; Mn^{II} formed in the presence of electron donors (EDTA, TEA, nicotine), Mn^{IV} in the presence of electron acceptors (e.g. methyl viologen)	[129]
$Mn^{III}(TSPc)(H_2O)(OH)^{4-}$ LMCT; IL(n, π^*)	$Mn^{II}(TSPc)^{4-}$	Solvent: water with adjusted pH; $\lambda_{irr} = 229$ or 254 nm; in the absence of 2-propanol as a radical scavenger, TSPc ring is decomposed; $\phi \sim 10^{-2}–10^{-3}$	[130]
$Mn^{III}(TPP)X$ $X = NO_3^-, NO_2^-, SO_4^{2-}, HSO_4^-$ $Mn^{III}(TPP)ClO_4$ mixed states IL + LMCT	$Mn^{II}(TPP)$ $Mn^{III}(TPP)Cl$	Solvent: frozen solvent glasses or polymer matrices; $\lambda_{irr} = 366$ nm; $\phi \sim 10^{-4}$; Q-bands ($\lambda > 500$ nm) are photochemically inactive; t = 10–298 K; $Fe(TPP)NO_3$ and $Fe(TPP)ClO_4$ are photostable	[131, 132]

$Fe^{III}(TPP)N_3$ $Mo^{V}O(TPP)N_3$	$Fe^{II}(TPP)$ $Mo^{IV}O(TPP)$	Solvent: benzene, toluene, CH_2Cl_2; $\lambda_{irr} > 400$ nm; N_3° evidenced by spin-trapping EPR	[133, 134]
$Fe^{III}(OEP)(MeIm)(CH_3OH)$ LMCT	$Fe^{II}(OEP)(MeIm)$	Solvent: CH_2Cl_2; λ_{irr} = visible light; both MeIm and CH_3OH are necessary for Fe(III) photoreduction (even in a trace amount)	[135, 136]
$Fe^{III}(OEP)L$ L = MeIm or 1,2-Me_2Im	$Fe^{II}(OEP)L$	Solvent: CH_2Cl_2; λ_{irr} = visible light; no Fe(III) photoreduction observed if L = Im, 1-MeIm, F^-, Cl^-, Br^-, I^-, or ClO_4^-	[137]
$Fe^{III}(TMP)$ X = Cl^- or OH^-	$Fe^{II}(TMP)$	Solvent: aromatic hydrocarbons; $\lambda_{irr} > 300$ nm; four-coordinated Fe(TMP) formed	[138]
$Fe^{III}(Por)Cl$ Por = TPP, TFPP, TTP, OEP, TMP, TDClP, Proto IX, LMCT states	$Fe^{II}(Por)$	Solvent: $CCl_4 + C_2H_5OH$; water + pyridine + alcohols; cyclohexene; polystryene matrices; $\lambda_{irr} > 350$ nm; $\phi \sim 10^{-2}$–10^{-3}; complexes catalyze the photoreduction of CCl_4 to Cl^- and CCl_3° by ethanol (turnover number $\sim 10^4$), and the oxidation of cyclohexene to 2-cyclo-hexene-1-ol and 2-cyclohexene-1-one	[139–145]
Fe(III) cytochrome c–N_3^-	Fe(II) cytochrome c	Solvent: water with adjusted pH; λ_{irr} = 340, 360, or 400 nm; $\phi \sim 10^{-4}$, wavelength and pH dependent; N_3° formed in the primary step	[146]
Fe(III) cytochrome c–$C_2O_4^{2-}$	Fe(II) cytochrome c	Solvent: water + methanol; λ_{irr} = 320–600 nm; $C_2O_4^{2-}$ bound on the surface of cytochrome cations; $\phi \sim 10^{-4}$	[147]
$Co^{III}(TPP)(O_2^-)$	$Co^{II}(TPP) + {}^3O_2$	Solvent: MeTHF; λ_{irr} = 532 nm; t = 140–200 K; fast recombination proceeds	[148]
$Co^{III}(Por)(AF)(O_2^-)$ Por = TPP, TAP, TTP, TFPP AF = $Fe(C_5H_5)(C_4NH_4)$	$Co^{II}(Por)(AF) + {}^3O_2$	Solvent: toluene; λ_{irr} = visible light; t = 100–200 K (azaferrocene AF coordinated by nitrogen atom)	[149]
$Co^{III}(TSPc)(H_2O)_{2-n}X_n^{(3+n)-}$ X = Cl^-, Br^-, I^-, N_3^- LMCT state	$Co^{II}(TSPc)^{4-}$	Solvent: water, pH ~ 1; $\lambda_{irr} > 350$ nm; $\phi \sim 10^{-1}$	[150]
$Co^{III}(TPPS)(CH_3OH)(OH)$	$Co^{II}(TPPS)(H_2O)$	Solvent: water + CH_3OH (or-2-propanol); λ_{irr} = UV + visible; Co(II) complex easily reoxidized by O_2; $\phi \sim 10^{-3}$	[151]

Table 4. (continued)

Complex, its photoreactive excited state	Products	Conditions and parameters of reactions Notes	Ref.
$Mo^{V}O(TPPS)X$ $X = OC_2H_5^-$ or NO_2^- IL(Soret) or LMCT states	$Mo^{IV}O(TPP)$	Solvent: toluene + ethanol; $\lambda_{irr} = 266$, 355 or 532 nm; $\phi = 0.03$ for ethoxycomplex at 532 nm; the lowest trip-quartet state is not photoredox reactive; the product reacts with O_2 or NO_2 giving $[MoO(TPP)]_2O$ or MoO(TPP)ONO, respectively	[152]
$Rh^{III}(OEP)(CH_3)$	$Rh^{II}(OEP)$	Solvent: benzene; $\lambda_{irr} = 255$–550 nm; $\phi \sim 10^{-2}$–10^{-1} and depends on λ_{irr}; Rh(OEP) either dimerizes to $[Rh(OEP)]_2$ or reacts with O_2 forming the adduct $Rh(OEP)O_2$	[153]
$Cr^{III}(TPP)(ONO)$ IL($^4S_1$ or $^4S_2$) state	$Cr^{IV}O(TPP) + NO$	Solvent: benzene; $\lambda_{irr} = 355$ nm or > 420 nm; $\phi = 0.1$ at 355 nm (CT band), 0.054 at 532 nm (Q-band), 0.11 at 447 nm (Soret band), and does not depend on the presence of O_2	[154]
$Cr^{III}(Por)N_3$ Por = OEP, TPP, TTP	$Cr^{V}N(Por) + N_2$	Solvent: toluene, benzene, MeTHF; UV + visible radiation; room temperature or 77 K; four complexes synthetized in 23–92% yield; isotopically labelled $Cr(TTP)^{15}N$ prepared from $Cr(TTP)^{15}N_3$	[125, 134, 155]
$Ti^{IV}(O_2^{2-})(TPP)$	$Ti^{IV}O(TPP) + {}^1O_2$	Solvent: benzene; $\lambda_{irr} > 330$ nm; $^1O_2$ formed exclusively from peroxo groups; in the primary step, $Ti^{II}(TTP)$ is formed which reacts subsequently with $Ti(O_2)(TTP)$	[157]
$Mn^{III}(TPP)(ONO_2)$ $Mn^{III}(TPP)(ONO)$ IL states (N, L, M-bands)	$Mn^{IV}O(TPP) + NO_2$ $Mn^{IV}O(TPP) + NO$	Solvent: benzene; $\lambda_{irr} = 350$–420 nm; $\phi = 1.6 \times 10^{-4}$ for $Mn(TPP)(NO_3)$, as an intermediate, $Mn(TPP)(NO_2)$ is formed; irradiation at $\lambda > 420$ nm does not induce the formation of MnO(TPP)	[132, 158]
$Mn^{III}(TPP)(XO_4)$ X = Cl or I	$Mn^{V}O(TPP)^+$	Solvent: hydrocarbons; $\lambda_{irr} = 310$–490 nm; $\phi = 2.7 \times 10^{-5}$; using ClO_4^-, the complex Mn(TPP)Cl is formed; the intermediate $MnO(TPP)^+$ behaves as an oxidant towards C_xH_y	[132, 159]
$Mn^{III}(Por)N_3$ Por = OEP, TMP, TPP, TTP	$Mn^{II}(Por)$ $Mn^{V}N(Por)$	Solvent: toluene, benzene, MeTHF; visible light; Mn(II) complex formed in MeTHF at room temperature, MeTHF oxidized to its cation radical; $Mn^{V}N(Por)$ diamagnetic complexes synthetized in 24–80% yield.	[124, 134, 160, 161, 156]

$Fe^{III}(TPP)(ONO_2)$	$[Fe(TPP)]_2O$	Solvent: benzene; $\lambda_{irr} = 350$–450 nm; as an intermediate, $Fe^{IV}O(TPP^+)$ is formed	[158]
$Fe^{III}(Por)N_3$ Por = TPP, OEP, TMsP	$Fe^VN(TPP)$	Solid complex irradiated; $\lambda_{irr} = 514.5$ or 488 nm; t = 30 K; a proof of $Fe \equiv N$ bond given by Raman spectroscopy (π-cation formation is ruled out); $[Fe(OEP)]_2N$ formed by local heating	[134, 162, 163]
$Sn^{IV}(Pc)_2$ short-lived CT state	$Sn^{II}(Pc) + Pc^0$	Solvent: halocarbons; $\lambda_{irr} = 640$ nm; recoordination of Pc^0 to Sn(Pc) involves the formation of $Sn^{III}(Pc)(Pc^+)$	[164]
$[Ru^{II}(OEP^{\circ})(CO)L]^+$ L = ethanol, py, Im MLCT(d, π)	$Ru^{III}(OEP)^+$	Solvent: CCl_4, $CHCl_3$, CH_2Cl_2, ACN; $\lambda_{irr} = 532$ nm, 355 nm, or unfiltered light; CO releasing proceeds the intramolecular electron transfer	[165, 166]
$Mn^{III}(Por)^{n+}$ Por = TPyP, TMPyP, TPPC, TPPS IL(π, π*) mixed with LMCT	$Mn^{IV}(Por)^{(n+1)+}$	Solvent: water with adjusted pH; $\lambda_{irr} = 465$–580 nm; $\phi \sim 10^{-2}$; photoreduction of various quinones sensitized by Mn(III) compounds	[167]
$Mn^{III}(TMPyP)^{5+} \ldots M^{II}(TSPc)^{4-}$ M = Cu or Ni	$Mn^{II}(TMPyP)^{4+}$ + $[M^{II}(TSPc^{\circ})]^{3-}$	Solvent: water (+ alcohols as scavengers); $\phi \sim 10^{-4}$, enhanced by aggregation; introducing O_2, reoxidation reactions occur	[168]
$Fe^{II}(Pc)L_2$ L = py, Im ^{3}IL(Q-band) state	$Fe^{III}(Pc)L^+$	Solvent: CH_2Cl_2 (+ CBr_4 as an electron acceptor); $\lambda_{irr} = 580$ nm; analogous Zn(II), Co(II) and Ru(II) are oxidized forming π-cation radicals	[169]
$Co^{II}(TPP)$	$Co^{III}(TPP)(O_2^-)$	Solvent: water; λ_{irr} = visible light; the product formed at a micellar interface (Triton X100)	[170]
$Co^{II}(OEP)$ IL(Q-band)	$Co^{III}(OEP)Cl$ $[Co^{III}(OEP^{\circ})Cl_2]^+$	Solvent: CH_2Cl_2 (+ CCl_4 as an electron acceptor); λ_{irr} = visible light; $Co^{III}(OEP^+)Cl_2$ formed at prolonged irradiation	[171]
$Os^{II}(TTP)(CO)L$ L = CH_3OH, py CTTS state	$Os^{IV}(TTP)Cl_2$	Solvent: CCl_4, $CHCl_3$ or CH_2Cl_2; $\lambda_{irr} = 333$, 365, 405 or 546 nm; $\phi \sim 10^{-1}$–10^{-2}, wavelength and solvent dependent; the lowest lying IL states are emitting, photochemically nonreactive states	[92]
$Os^{II}(OEP)[P(OCH_3)_3]_2$ ^{3}MLCT	$Os^{IV}(OEP)Cl_2$	Solvent: CCl_4, $CHCl_3$ or CH_2Cl_2; $\lambda_{irr} = 365$ or 405 nm; $\phi = 0.004$ in CH_2Cl_2 and 1.4 in CCl_4; radical species involved in the product formation	[192]

complexes undergo such a photoelimination, e.g. [149]

$$Co^{III}(Por)(AF)(O_2^-) \xrightarrow{h\nu} Co^{II}(Por)(AF) + O_2 \quad (24)$$

where the axial ligand azaferrocene $Fe(C_5H_5)(C_4NH_4)$, (denoted AF) is coordinated to the cobalt central atom by a nitrogen atom of the cyclopentadienyl-like ring C_4NH_4 in which one of the CH fragments is substituted by the nitrogen atom.

Photoredox additions and eliminations are limited to complexes with a few axial ligands, molecular O_2 being the only one which is significant. As a product of photoeliminations, O_2 can be formed in its triplet ground state (as an example can serve the above-mentioned elimination from the superoxide $Co^{III}(Por)(AF)(O_2^{-1})$ or in an excited singlet state, e.g. [157]

$$2Ti(TPP)(O_2^{2-}) \xrightarrow{h\nu} 2TiO(TPP) + {}^1O_2 \quad (25)$$

Processes of axial ligand photooxidation lead to the production of radicals which are consumed in subsequent fast thermal (dark) reactions. Such radicals can dimerize, as e.g. alkyl radicals released from excited Co(III) organometallics [122]

$$Co^{III}(TP)(CH_3) \xrightarrow{h\nu} Co^{II}(TP) + CH_3^{\bullet} \quad (26)$$

$$2CH_3^{\bullet} \longrightarrow C_2H_6 \quad (27)$$

or they can recoordinate to the reduced central atom

$$Co^{II}(TP) + CH_3^{\bullet} \longrightarrow Co^{III}(TP)(CH_3) \quad (28)$$

or react with surrounding molecules (solvent molecules), as e.g. atomic chlorine $Cl^{\bullet}$ released from excited $Fe^{III}(TP)Cl$ in hydrocarbon solvents RH [138]

$$Fe^{III}(TP)Cl \xrightarrow{h\nu} Fe^{II}(TP) + Cl^{\bullet} \quad (29)$$

$$Cl^{\bullet} + RH \xrightarrow{h\nu} Cl^- + H^+ + R^{\bullet} \quad (30)$$

It should be pointed out that the simultaneous occurrence of all three sorts of radical reaction (expressed by Eqs. 27, 28, 30) cannot be excluded.

A photoreduction of an axial ligand associated with the central atom photooxidation is possible in principle (MLCT states might be responsible for such a process); it is not, however, a common photochemical process.

An electron transfer from a coordinated axial ligand L to the central atom M, followed by the splitting of the M–L bond, is a well-documented process for organometallics and a suggested mode of deactivation for a number of complexes containing inorganic anionic ligands (e.g. halogenides) or alkoxides. It is generally accepted that an LMCT excited state (where L means axial ligand) is responsible for the process. In the photochemistry of metallotetrapyrroles, where the absorption bands of LMCT transitions are often hidden under the envelope of porphyrin-localized intraligand bands, the relationship LMCT state $\longleftrightarrow$ photooxidation of an axial ligand cannot be clarified so unambigously. This seems to be a reason of why discussion of this relationship is frequently omitted in published papers.

In all available papers dealing with the photooxidation of axial ligands it is tacitly assumed that tetrapyrrole rings remain apparently chemically unchanged. Such uninvolvement of equatorial ligands has also been observed for complexes possessing other quadridentate than tetrapyrrole ligands (e.g. open-chain Schiff base ligands [29, 172–175], and saturated macrocyclic ligands [176–178].

An interesting, photocatalytically promising type of photoredox reaction is the processes belonging to class (b) of the above classification. The term atom-transfer reactions (which is not fully precise) is used for them. For ground-state chemistry such reactions are very uncommon. As a rule, the central atom oxidation number is increased in those reactions in two (e.g. $Mn^{III} \rightarrow Mn^{V}$) or one (e.g. $Mn^{III} \rightarrow Mn^{IV}$) unit. The transferred atom (or remaining atom having been previously coordinated to the central atom) is in the cases known so far either nitrogen N^{-III} or oxygen O^{-II}, e.g. [134, 132]

$$Mn^{III}(Por)N_3 \xrightarrow{h\nu} Mn^{V}N(Por) + N_2 \tag{31}$$

$$Mn^{III}(Por)(ONO) \xrightarrow{h\nu} Mn^{IV}O(Por) + NO \tag{32}$$

Very detailed studies on isotopically labelled complexes prove that the remaining atoms (N, O) have their origin in formerly anionic ligands (N_3^-, NO_2^-, NO_3^-, ClO_4^-, IO_4^-). Furthermore, it has been shown that the bond order values in the nitrido (M≡N) and oxo (M=O) complexes are 3 and 2, respectively, i.e. the "octet rule obeying" oxidation states of nitrogen and oxygen atoms corresponds to the reality. The oxo complexes behave as strong oxidants towards organic compounds. Along with the study of various theoretical aspects and reaction mechanisms, connected with the formation and reactivity of oxo and nitrido complexes, photochemistry was used also as a tool for the synthesis of such coordination compounds [134, 155, 156].

It should be pointed out that the nature of the primary photochemical step(s) is still obscured and can depend, even for the same complex, on experimental conditions. Thus, $Fe^{III}(Por)N_3$ converts under irradiation in the solid state at low temperatures [162, 163] into $Fe^{V}N(Por)$; in some solution systems [133] the formation of azidoradicals $N_3^{\bullet}$ has been detected by spin-trapping EPR; no information on the heterolytic splitting of the Fe–N_3 bond yielding N_3^- anion has been described in the literature (for azido complexes of some other central atoms the photosubstitution of the coordinated N_3^- ligand is a dominant chemical deactivation mode [1]). In addition, at particular conditions, the photolysis of $Fe^{III}(Por)N_3$ produces μ-nitrido binuclear mixed-valence complexes, e.g. $(TTP)Fe^{III}–N–Fe^{IV}(TTP)$ [134].

Research into oxo and nitrido complexes has been encouraged by the finding that the formation of ferrylporphyrins $Fe^{IV}O(Por)$ plays a key role in the reaction cycle of cytochrome P-450, as well as by knowledge of intramolecular and intermolecular nitrogen atom transfer and its incorporation into C–H bonds, catalyzed by cytochrome P-450 derivatives, e.g. isozyme P-450-LM3,4 isolated from rabbit liver [179]. Synthetic nitrido and oxo iron porphyrins may be regarded as models of intermediate species involved in biological nitrogen and oxygen atom transfer processes.

The third class (c) of photoredox processes, namely electron transfer from the coordinated tetrapyrrole ring (its π-system) to the central atom or vice versa, comprises a few cases, represented in Table 4 by the neutral complex $Sn(Pc)_2$ [164] and thermally stable π-cation radicals $[Ru(OEP)(CO)L]^+$ [165]. In the former case a molecular phthalocyanine Pc is formed under irradiation. It was suggested that the photoreactive excited state was a short-lived charge-transfer state reached from the phthalocyanine ring localized triplet $^3(\pi, \pi^*)$. The molecular Pc recoordinates to the central atom Sn^{II} forming the complex $Sn^{III}(Pc^{2-})(Pc^-)$ in the first step. The complex of Sn^{III} then converts to the parent compound $Sn(Pc)_2$.

The formation of some Ru(III) complexes upon photodecarbonylation followed by an internal electron transfer can be expressed as follows

$$[Ru^{II}(OEP^{\bullet})(CO)L]^+ \xrightarrow{h\nu} [Ru^{III}(OEP)L]^+ + CO \tag{33}$$

Electrochemically produced Ru(II) porphyrin π-cation radicals $[Ru(OEP^{\bullet})(CO)L$, where $L = C_2H_5OH$, py, Im or Br^- are thermally stable in their $^2A_{2g}$ or $^2A_{1g}$ ground states. Electrochemical [166] and photochemical [165] investigation of the preparation, stability and reactivity of the complexes brought evidence that the internal electron (hole) transfer depends on the electronic properties of the axial ligands. When a CO molecule (a strong π acceptor) is coordinated as an axial ligand, Ru(II) π-cation radicals are more stable than the corresponding Ru(III) complexes. In other cases (e.g. when some phosphines are bonded to the central atom), Ru(III) complexes are stabilized. Picosecond transient absorption spectrometry revealed that the internal electron transfer was the second reaction step, the first one being the photoelimination of a molecule CO from the primary coordination sphere of a Ru(II) π-cation.

It cannot be excluded that for first-row transition metal complexes, in addition to the above factors, the central atom spin state also plays a significant role in the stability of individual electronic and/or spin isomers, and in their ability to undergo photochemical interconversions. For some iron complexes such spin-state photoisomerizations are well-known [1, 121].

Long-range electron-transfer reactions have been investigated in two main groups of metallotetrapyrroles. Natural and synthetic Mg(II) and Zn(II) complexes, belonging to the first group, have been studied in connection with mimicking electron transfer in various stages of photosynthesis. These processes are briefly mentioned in Sect. 7.

The second group comprises cytochrome c or cytochrome-like and hemoglobin-like compounds in which the starting and ending site of the electron-transfer processes is the iron central atom able both to accept an electron (Fe(III) → Fe(II)) or to donate it (Fe(II) → Fe(III)). The biological function of cytochromes does not origin in photochemical reactions. On the other hand, only ultrafast techniques (the majority of them being photochemical and photophysical) can provide relevant answers to the questions concerning the mechanism of cytochrome biological action. The studies in this field are focused mainly on three problems:

1) channels for electron entering and leaving, their connection to the protein structure and "electrical conductivity" of the tetrapyrrole π-system;
2) environmental effects, including the incorporation of cytochromes in micelles, bilayers, and the separation of reaction sites by membranes;
3) relations between the rate constant of electron-transfer processes, driving forces, distance and mutual orientation of reaction sites.

The results obtained are given and reviewed elsewhere [9–14, 180–187]. They are beyond the scope of this Chapter as they deal predominantly with natural systems.

Outer-sphere electron-transfer reactions are, in general, the best elaborated redox processes from the theoretical point of view. Expansion of activity in the study of such reactions has been motivated chiefly by the aim of preparing systems able to convert solar energy into chemical and electrical energies with high efficiency, and prepare thus energy in an economically profitable and ecologically pure mode. It seems to be worth mentioning that whereas the rate constant of a photoinduced electron transfer (an excited state parameter) can be correlated with some ground-state parameters of complexes (e.g. redox potential values) and thus optimized, the efficiency of overall photoredox processes, expressed through its quantum yield value, Φ, cannot be correlated with the ground-state parameters of complexes or their parts (the central atom and ligands). Current knowledge, therefore, does not allow us to utilize known ground-state characteristics of complexes for modelling their composition and structure in order to obtain just the required (i.e. as highest as possible for solar energy conversion processes) efficiency of their photoredox reactions [29].

There is a lot of examples of outer-sphere electron-transfer reactions occurring in irradiated systems of typical inorganic complexes [188–191]. However, for metallotetrapyrroles such reactions involving the central atom are rather rare. It should be underlined that it is usually not so simple to distinguish between the primary outer-sphere and inner-sphere step, particularly in cases when both lead to the same product and proceed with comparable rates. Moreover, a number of outer-sphere electron-transfer reactions occur as reversible processes with no net chemical change. To solve this problem, techniques of the incorporation of a reactant into bilayers or micelles and its separation from a reaction partner, which was successfully used in differing outer-sphere and inner-sphere modes of an electron transfer in systems containing cytochrome c [147] and its models [179], seem to be promising.

Discussing photoredox reactions, the term "ligand aided processes" and "secondary sphere ligand participating processes" must be distinguished. The former term describes such reactions for which the presence of a certain ligand in the primary coordination sphere is a necessary condition, the ligand itself, however, does not undergo a chemical change. Freely speaking, the effect of such a ligand can be characterized as catalytic. The latter term is associated with processes in which a molecule of the secondary coordination sphere behaves as the redox partner of an excited complex but does not enter its primary coordination sphere. A typical example of ligand aided process is the photoreduction of Mn(III) to Mn(II) in various porphyrin and phthalocyanine complexes [128] occurring only when a pyridine molecule is coordinated in axial position (pyridine itself does not participate directly in the reaction). An electron transfer from an anion $C_2O_4^{2-}$ located on the surface of the protein part of cytochrome c to the central atom Fe(III) can serve as an example of a "secondary sphere ligand participating process" [147].

Photooxidation of the central atom Os(II) in hexacoordinated porphyrin complexes is supposed to start with the ejection of an electron from an charge-transfer to solvent excited state, CTTS, of the complexes. A complicated set of elimination, addition and redox steps involving radicals terminates in the formation of the complexes $Os^{IV}(Por)Cl_2$. Solvent molecules (CCl_4, $CHCl_3$, CH_2Cl_2) served as a source of chlorine atoms [92, 192].

Outer-sphere photoredox reactions are often interpreted as a consequence of ion-pair charge-transfer, IPCT [168] or charge-transfer to solvent, CTTS [92] excited states. In principle, however, any kind of excited state can be involved in such processes.

7 Tetrapyrrole Ring Localized Photoredox Reactions

Inorganic photochemists used to focus their attention on the changes involving the central atom. In most studies of tetrapyrrole complexes it is tacitly (and justifiably) supposed that their macrocyclic ligand remains untouched in chemical reactions. On the other hand, there are several biochemical processes concerning the formation, function, and degradation of tetrapyrrole biocomplexes "in vivo" which are associated and conditioned by changes in the composition of their rings. Some of them, e.g. the fate of the chlorophylls in senescent chloroplasts or the catabolism of heme complexes in vertebrates still belong to the category of chemical and biological enigmas. Understanding such processes is a challenge for chemists in general and photochemists in particular.

This Section deals with the photoinduced reactions in which the central atom oxidation state is preserved and just a tetrapyrrole ligand undergoes a chemical change yielding stable or intermediate products. Examples are compiled in Table 5. Ring localized photoprocesses are photoredox reactions which can be classified into five groups:

1) π-cation and π-anion formation by photoinduced electron transfer processes;
2) cyclic to linear tetrapyrrole ring-opening;
3) photooxidations in which the carbon–carbon sequence is interrupted and a new "heterocyclic element" is introduced into the ring;
4) photooxidations accompanied by an increase in ring double bond number (e.g. chlorin-to-porphyrin reactions) and photoreductions accompanied by a decrease in ring double bond number (e.g. porphyrin-to-chlorin conversions);
5) processes localized on the peripheral substituents of the ring.

The course of the photoformation of the tetrapyrrole ring-localized radicals (group 1 of the above classification) can be expressed by four equations

$$*M(TP) + A \rightarrow M(TP^{\bullet})^{+} + A^{-} \tag{34}$$

$$*M(TP) + D \rightarrow M(TP^{\bullet})^{-} + D^{+} \tag{35}$$

$$*M(Pc) + SH \rightarrow M(Pc - H^{\bullet}) + S^{\bullet} \tag{36}$$

$$*R - M(TP) \rightarrow M(TP^{\bullet})^{-} + R^{\bullet} \tag{37}$$

(A = electron acceptor, D = electron donor, SH = molecule of a protic solvent, R = axial alkyl anionic ligand).

The first two processes are common reactions for complexes in their intraligand singlet or triplet (π, π*) excited states. Since excited molecules are both better reductants and better oxidants than those in the ground state [1], the same complex can be oxidized to its π-cation radical (when quinones, nitroaromatics, halocarbons, viologenes or other oxidizing agents are present) or reduced to its π-anion radical (in the presence of a reductant, e.g. NAHD or L-ascorbic acid). The radicals formed are usually unstable species; the anion radical $[SbO(UroP^{\bullet})]^{6-}$ formed at a photolysis of $SbO(UroP)^{5-}$ in alkaline solutions is one of the rare exceptions [216].

Back electron-transfer processes of π-anion and π-cation radicals with reversible electron donors or acceptors (e.g. aquated Fe^{3+}, $[Fe(CN)_6]^{3-}$, quinones) are fast reactions realized in nano- or picosecond time scale. In cases when irreversible redox partners are used (e.g. $S_2O_8^{2-}$, CBr_4, CCl_4, EDTA) tetrapyrrole ring ring localized radicals dimerize [193], decompose [212], undergo disproportionation [215] or other stabilization reactions. Photoformation of stable products will be discussed later.

Hydrogen-atom abstractions were observed in the systems containing some metallophthalocyanines excited in the Soret band by ultraviolet radiation [217].

The last class of the ring localized radical photoformations, the production of zwitterionic products associated with axial ligand oxidation, was found to occur in systems containing complexes of non-transition metals having stable (and only one) oxidation state, such as Ge^{IV}, Al^{III}, and In^{III}. The photoreactive excited state can be characterized as a mixing of an intraligand (π, π*) and

Table 5. Tetrapyrrole ring localized electron-transfer and atom-transfer photoredox reactions

Complex-reactant, its excited state	Quencher	Complex-product	Conditions, reaction parameters, notes	Ref.
$Mg^{II}(Pc)L_2$ L = Im, CN^-, MeIm, py, Mepy	CBr_4	$[Mg^{II}(Pc^{\cdot})L_2]^+$	Solvent: CH_2Cl_2 or $C_2H_4Cl_2$; $\lambda_{irr} = 670$ nm; prolonged irradiation causes decomposition of the ring; the product $Mg(Pc^+)L_2$ dimerizes	[193]
$M^{II}(TPP)$ M = Mg, Zn, H_2	chloroalkanes quinones	$[M^{II}(TPP^{\cdot})]^+$	Solvents: CCl_4, CH_2Cl_2 or $C_2H_2Cl_4$; $\lambda_{irr} = 417$ nm; $\phi = 0.79$ for Mg(TPP), 0.16 for Zn(TPP), $< 10^{-4}$ for Cu(TPP)	[194]
Mg(TMOP)–R–H_2(TMOP) R = bridge of cofacial porphyrins		$[Mg(TMOP^{\cdot})]^+$–R–$[H_2(TMOP^{\cdot})]^-$	Solvent: CH_2Cl_2; $\lambda_{irr} = 527$ nm; = 1; 90% of the singlet state energy is stored in the photoproduct; fast back electron transfer	[195]
Mg(DOP)–$(R)_2$–H_2(DOP)		$[Mg(DOP^{\cdot})^+$–$(R)_2$–$[H_2(DOP^{\cdot})]^-$	Solvent: acetone, CH_2Cl_2, DMF or alkylacetates; $\lambda_{irr} = 532$ or 588 nm; the charge-recombination rate constant correlates with the reverse of the solvent relaxation times	[196]
Zn^{II}(PCOO), $^1(\pi, \pi^*)$	dinitrobenzoic acid	$[Zn^{II}(PCOO^{\cdot})]^+$	Solvent: CH_2Cl_2; $\lambda_{irr} = 580$ nm; electron transfer mediated by a hydrogen-bonded interface has $k = 3 \times 10^{10}\ s^{-1}$	[197]
Zn(TPP), $^3(\pi, \pi^*)$	BQ	$[Zn(TPP^{\cdot})]^+$	Solvent: 4-methylpentan-2-one; $\lambda_{irr} = 532$ nm; $\phi \sim 10^{-1}$, enhanced by an addition of neutral salts	[198]
Zn(OEPC), $^3(\pi, \pi^*)$	Duroquinone	$[Zn(OEPC^{\cdot})]^+$	Solvent: ethanol or liquid crystals E-7; $\lambda_{irr} = 580$ nm; fast back electron transfer	[199]
$Zn(Por)^{n+}$, $^3(\pi, \pi^*)$ Por = TMPyP, TPPS or TAFP	MV	$[Zn(Por^{\cdot})]^{(n+1)}$	Solvent: water; λ_{irr} = white light; complexes covalently bound to various polymers or copolymers	[200]
Zn(TOOPP) (py), $^1(\pi, \pi^*)$	RuO_2	$[Zn(TOOPP^{\cdot})]^+$	Solvent: water/oil microemulsion containing SDS or DTAB, THF, pyridine; $\lambda_{irr} = 532$ nm; the complex is covalently attached through a flexible chain of varying lenght to solid RuO_2	[201]

$Zn(TMPyP)^{4+}$, $^3(\pi, \pi^*)$	$Fe^{3+}(aq)$ $Co^{III}(EDTA)^-$ $S_2O_8^{2-}$	$[Zn(TMPyP^{\cdot})]^{5+}$	Solvent: water; λ_{irr} = visible light; in micellar systems the back electron transfer is retarded	[202, 203]
Zn(Pc)L L = Im, py, CN^- (π, π^*–Q–band)	CBr_4	$[Zn(Pc^{\cdot})L]^+$	Solvent: CH_2Cl_2 or pyridine; $\lambda_{irr} = 580$ nm; $\phi \sim 10^{-2}$–10^{-3}	[169]
$Zn(TAPP)^{4+}$ $^1(\pi, \pi^*)$	$Cu^{II}(TPPS)^{4-}$	$[Zn(TAPP^{\cdot})]^{5+}$ $[Cu(TPPS^{\cdot})]^{5-}$	Solvent: acetone + water, or DMSO + glycerol; $\lambda_{irr} = 347$ or 560 nm; ion pairs formed both in solutions and glycerol glasses at 77 K; only the excited singlet of Zn(II) complex acts as an electron donor	[204]
Zn(Por)–Au^{III}(Por′) Por = TPPC, TPPS, TAPP Por′ = TPPC, TPPS, TMPyP		$[Zn(Por^{\cdot})]^+ \cdot [Au^{III}(Por'^{\cdot})]^-$ (contact ion pair dimer)	Solvent: water; $\lambda_{irr} = 532$ nm; the electron transfer occurs within the face-to-face dimers held by interactions of π and π^* orbitals; fast charge recombination	[205]
Zn(BP)–phen–Au^{III}(BP)		$[Zn(BP^{\cdot})]^+$–phen–$[Au(BP^{\cdot})]^-$	Solvent: DMF; $\lambda_{irr} = 532$ or 598 nm; $\phi \sim 1$; both parts of the dimer can be excited separately; in electron transfer reactions triplet and singlet Zn(II) and triplet Au(III) take part	[206]
$M_a^{II}(TPPC)^{4-}$–$M_b^{II}(TMPyP)^{4+}$ M_a = Zn, Pd, H_2 M_b = Zn, Cd, Pd, H_2		$[M_a(TPPC^{\cdot})]^{3-}$–$[M_b(TMPyP^{\cdot})]^{3+}$	Solvent: water + methanol; $\lambda_{irr} = 580$ or 600 nm; very fast charge recombination ($k \sim 10^9$–10^{10} s^{-1}); the charge separation is adiabatic, activationless and solvent-controlled electron transfer	[207]
$Ge^{IV}(TPP)R_2$ M^{III}(Por)R M = Al or In: R = alkyl Por = TPP or OEP		$[Ge^{IV}(TPP^{\cdot})R]^-$ $[M^{III}(Por^{\cdot})]^-$	Solvent: THF, DMF or benzene; $\lambda_{irr} = 580$, 595 or 450–600 nm; ϕ is solvent dependent, $\sim 10^{-1}$–10^{-3}; radicals R° and zwitterionic complexes formed	[208, 209]
Co^{III}(OEP)X X = Cl^-, Br^- or ClO_4^-	CCl_4 or $FeCl_3$	$[Co^{III}(OEP^{\cdot})X_2]^+$	Solvent: $CH_2Cl_2 + CCl_4$; λ_{irr} = visible light; Co^{III}(OEP)X produced by photooxidation of Co^{II}(OEP)	[171, 210]
Zn^{II}(Mb)	$[Fe(CN)_6]^{3-}$	$[Zn^{II}(Mb^{\cdot})]^+$	Solvent: water; $\lambda_{irr} > 450$ nm; both dynamic and static quenching of the triplet Zn(II) complex occur; very fast back electron transfer	[211]

Table 5. (continued)

Complex-reactant, its excited state	Quencher	Complex-product	Conditions, reaction parameters, notes	Ref.
$M(TSPc)^{n-}$ M = Co^{II}, Co^{III}, Cu^{II} or Ni^{II}		$[MTSPc^{\cdot})]^{1-n}$	Solvent: water; λ_{irr} = visible light; cation radicals can decompose; $Co^{II}(TSPc^{+})$ converts to $Co^{III}(TSPc)$	[212]
$Cu^{II}(TP)$, (n, π^*) TP = TSPc or TOPc	Propanol	$(Cu^{II}(TP{-}H)]$	Solvent: $CHCl_3$ containing hydrogen atom donor; $\lambda_{irr} = 280, 300, 320$ or 350 nm; Cu(I) and Cu(III) complexes do not transform into Cu(II) ligand radicals; $\phi \sim 10^{-2}$–10^{-3}	[213]
$Ru^{II}(Pc)LL'$, $^3(\pi, \pi^*)$ L, L′ = py, py; DMF, CO; Mepy, CO; or DMSO, DMSO	CBr_4 quinones nitroaromatics Fe^{3+}	$[Ru^{II}(Pc^{\cdot})LL']^{+}$	Solvent: CH_2Cl_2 or MeTHF at 79 K; $\lambda_{irr} > 580$ nm (Q-band); π-cations are stable species; the formation of exciplexes observed	[56, 214]
$Lu^{III}(TMPyP)^{5+}$,	NAHD $S_2O_8^{2-}$	$[Lu^{III}(TMPyP^{\cdot})]^{4+}$ $[Lu^{III}(TMPyP^{\cdot})]^{6+}$	Solvent: water; $\lambda_{irr} = 532$ nm; the cation in N_2-atmosphere disproportionates to $Lu^{III}(TMPyP)^{5+}$ and $Lu^{III}(TMPyP)^{3+}$, in O_2-atmosphere O_2^- is formed	[215]

LMCT states. Unlike bimolecular, outer-sphere, hydrogen-atom abstraction reactions, the formation of zwitterionic products is a typical innercomplex process.

Intramolecular light-driven electron-transfer processes have been followed in many systems containing heterodimers (in which the monomer units are bound via π-orbital interaction of their two pyrrole rings in a face-to-face manner [205] or through a linking group [206], as well as in the systems containing dyads D–A, triads (D–D–A, D–A–D, D–A–A, or A–D–A) or even more complicated structures, in which the donors D and acceptors A are covalently linked, one of them being a metallotetrapyrrole. Such systems are both the functional and structural models of the naturally occurring chlorophylls and bacteriochlorophylls. They are, however, easier to make and handle than natural dimers present in photosynthetic reaction centers (there are some differences, e.g. the bacteriochlorophyll dimers have a slipped face-to-face structure with the overlap of a single pyrrole ring of each monomeric unit [207].

Photosynthetic model systems have recently been exhaustively reviewed elsewhere [5, 6, 218] and a number of results are given in the latest literature [219–224]. The attention of the researchers is focused on topics such as electron-transfer chain and energy dissipation within models (the first step is the transfer of an electron from a metallotetrapyrrole moiety yielding a cation radical); the dependences of the electron-transfer rate constant on the driving force of the process; distance and mutual orientation of donor and acceptor sites; influences of membranes and medium (solvent) properties, etc.

Discussion of photochemical properties of natural and artificial photosynthetic centers is beyond the scope of this chapter. This part of photochemistry of tetrapyrrole complexes involving systems seems to be the most elegant part in the field and one of the most exciting parts in photochemistry and photobiology as a whole.

Photooxidative ring cleavage reactions, probably playing a role in the degradation of tetrapyrrole biocomplexes, have been investigated for several vitamin B_{12} analogues [122, 225–230]. The formation of one isomeric dioxosecocorrin (two isomers are formed) by a regioselective photooxygenolytic cleavage of the corrin ring can be schematically expressed as follows (for the sake of simplicity, in the following reactions the substituents and axial ligands are not specified)

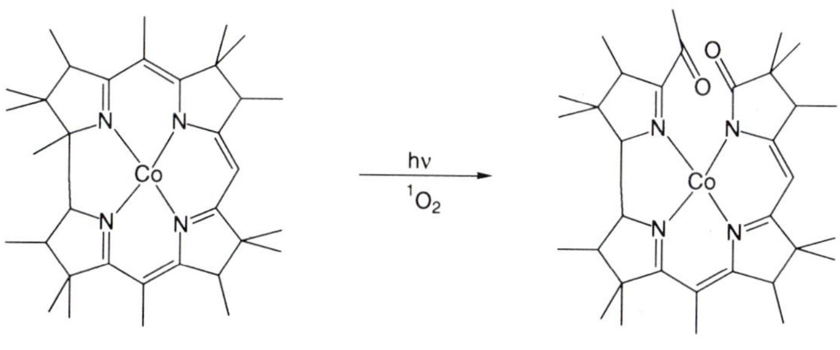

(38)

The reaction mechanism involves as attack of electrophilic singlet oxygen $^1O_2$ on the corrinoid π-system. $^1O_2$ is generated by the energy transfer from either a photoexcited complex or a sensitizer (methylene blue). Vitamin B_{12} itself is not photosensitive. The yield of the products depends on ring substituents, in all cases, however, the reaction is very selective.

In addition to the regioselective ring-opening photoreactions, photolysis can lead to a deep degradation of tetrapyrrole ligands which was suggested, e.g. as a mode of photodecomposition of the complex Cu(ODSPc) [213].

The third group (class 3 of the above classification) of the tetrapyrrole ring localized photoreactions can be exemplified [231] by the formation of oxonia-chlorins from chlorophyll derivatives (X = Cl, CF_3COO)

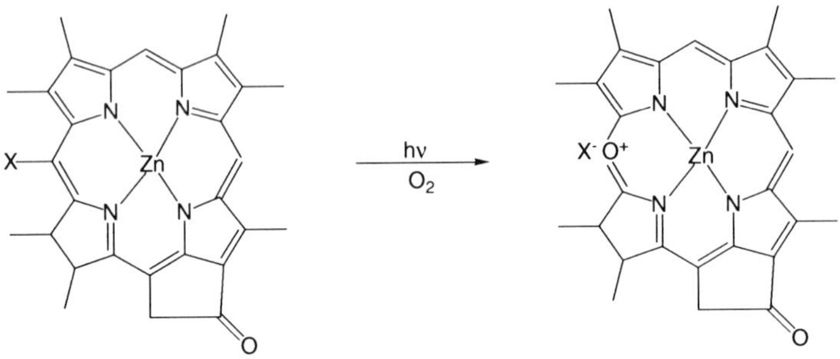

(39)

Similarly to the vitamin B_{12} derivatives discussed above, the photoinduced cleavage of the tetrapyrrole ligands, loss of a carbon atom and final product formation, are highly regioselective processes, too. Analogous reactions also provide some magnesium and zinc chlorins and porphyrins [232–234] (e.g. Mg(OEP) is photooxygenated in dry benzene in the presence of air to 4-formyl-5-oxa-octaethylporphinato magnesium(II)). Oxonium derivatives are precursors for preparation of acyclic tetrapyrrole compounds (e.g. 4-formyl-5-oxa-octaethylporphinato Mg(II) is very labile to hydrolysis and it is converted to the metal-free 1,γ-dioxo-8′-formyl-octaethyl violin [234]).

Electron transfer photooxidation of tetrapyrrole ligands proceeds usually via π-cation (π-dication) intermediates formed in the primary photochemical step (class 1 reactions). π-cations being strong electrophiles are able to bind nucleophilic species producing thus rings with a lower number of double bonds. As an example the photoformation of dihydroxysubstituted tetrapyrrole Zn{TPPS – $(OH)_2$} from the corresponding complex Zn(TPPS) is presented [235, 236]. The reaction occurs in the presence of oxygen, at pH = 12, using incident light with λ_{irr} = 422 nm (R = $C_6H_4SO_3^-$):

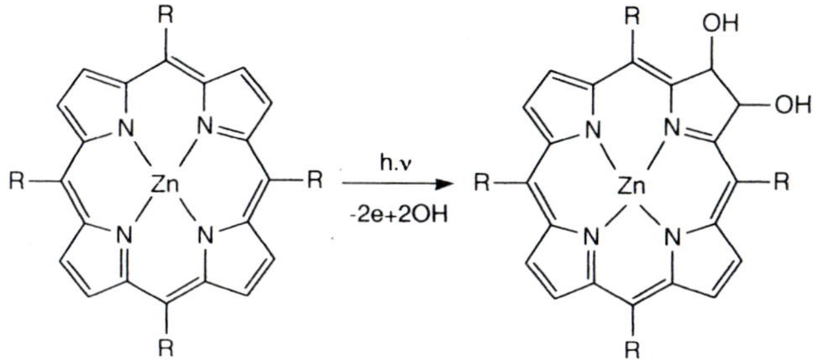

(40)

A high selectivity of the product formation is due to the inclusion of the reactant in β-cyclodextrin cavities.

It was shown that the formation of the doubly oxidized 1,2-dihydroxyporphyrin Zn{TPPS – $(OH)_2$} is not the last step of photochemical ring localized transformations. Prolonged photolysis causes the formation of the complex Zn{TPPS – $(OH)_4$}. In all the above mentioned hydroxyporphyrins the axial positions are occupied by OH^- and/or H_2O ligands depending on the pH value of the irradiated solutions.

Since the pioneering work of Krasnovskii [237] the photoreduction of metalloporphyrins and metallochlorins (class 5 reactions) has attracted the interest of many researchers because of its possible relationship to the biosynthetic formation of chlorophylls and related processes. Most of the work deals with complexes of redox stable central atoms and the primary photochemical step is, therefore, the transfer of an electron from a reductant (e.g. L-ascorbic acid) to a tetrapyrrole forming a π-anion radical. Subsequent reactions, occurring also in the absence of light, lead to final products. In such reactions porphyrins P (having 11 ring double bonds) are reduced to chlorins or phlorins H_2P (isomeric forms of H_2P possessing 10 ring double bonds), subsequently to bacteriochlorins or isobacteriochlorins H_4P (9 ring double bonds) or even more reduced forms (e.g. to porphyrinogens H_6P with 8 ring double bonds).

A very instructive example of stepwise photoreduction: porphyrin → chlorin → isobacteriochlorin is provided by tin(IV) porphyrins [238, 239], e.g. $Sn(OEP)^{2+}$ is photoreduced in the presence of $SnCl_2 \cdot 2H_2O$ (in which Sn(II) acts as an electron donor and H_2O as a proton donor) in degassed pyridine to tin(IV) octaethylchlorin and subsequently to tin(IV) octaethylisobacteriochlorin The use of $SnCl_2 \cdot 2D_2O$ led to the formation of deuteriated reduced products.

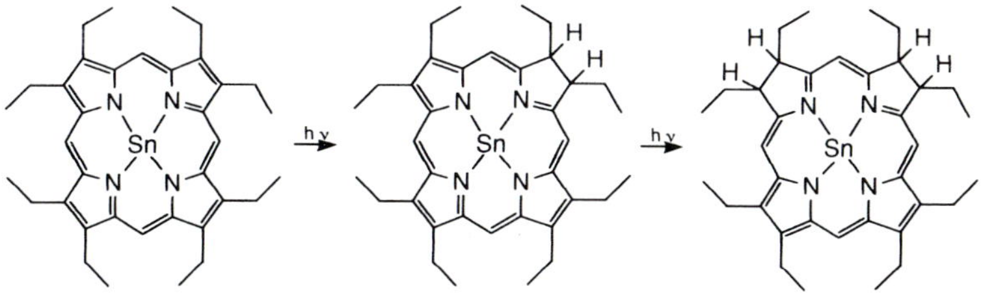

(41)

Irradiation of tin(IV) chlorin or bacteriochlorin in the presence of oxygen gave rise to the formation of the parent $Sn(OEP)^{2+}$.

A general procedure for regioselective synthesis of chlorins from porphyrins is described in Ref. [240]. The method is based on the experimental evidence that certain substituents (e.g. vinyl, acetyl, formyl) direct the photoreduction to the site of the ring to which they are attached. As an example, the photochemical synthesis of zinc(II) vinylchlorins from the corresponding porphyrins can be demonstrated (R = H, CH_3, C_2H_5, $C_2H_4COOCH_3$)

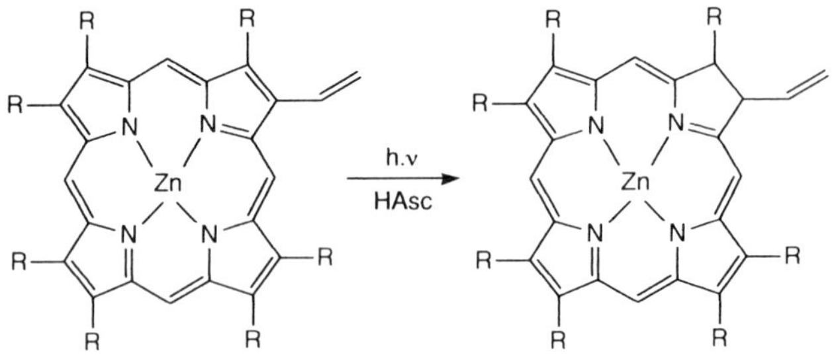

(42)

In the case of an acrylic derivative, the primary reduction takes place at the porphyrin ring. The migration of the acrylic double bond to the reduced ring rapidly follows and the final product of the photoreduction is zinc(II) porphyrin propionate. The mechanism of the reactions can be schematized as follows

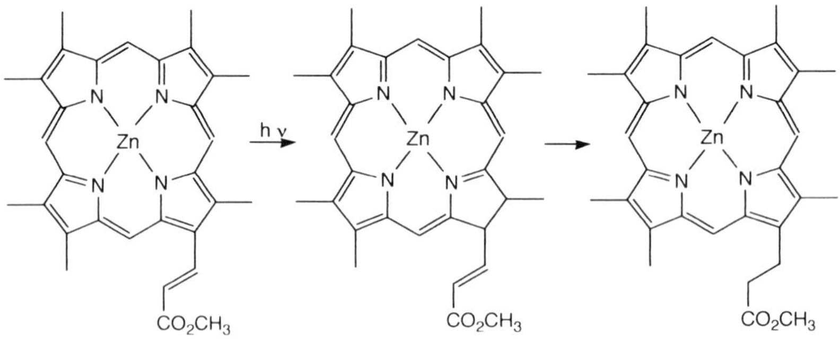

(43)

A migration of a double bond, localized on a remote group, was observed to occur in the course of the photoreduction of metallochlorins to isobacteriochlorins [241, 242]. The process can be expressed as follows

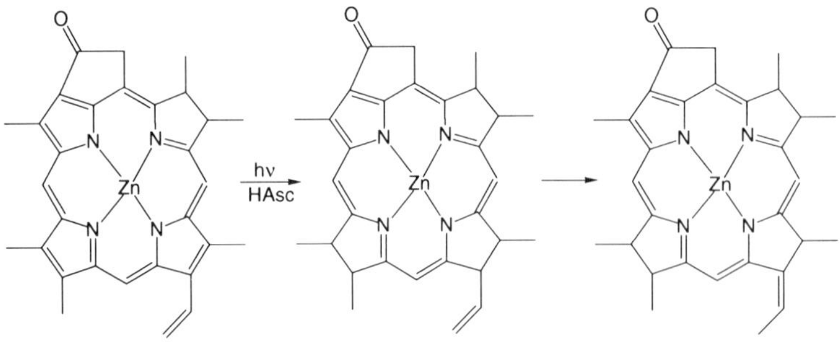

(44)

A ring localized hydrogenation can be a consequence of the disproportionation of photochemically formed π-anion radicals. Such a process was found [243] to proceed in aqueous solutions containing two gold(III) porphyrins Au(Por)Cl, where Por = TMPyP or TPyP, and electron donors (EDTA or TEA). π-radicals abstract protons from H_3O^+ ions (pH < 4) and give the parent porphyrin complexes Au(Por)Cl and reduced phlorins $Au(H_2Por)Cl$.

The last group of the tetrapyrrole ligand localized photochemical changes (class 6 reactions) concerns processes on the peripheral ring substituents. For inorganic photochemists those reactions do not belong to the most interesting ones. They may, however, play a crucial role in photochemical "in vivo" reactions (e.g. the processes localized at the protein chains). The reactions can be exemplified by an ether formation on prolonged irradiation of an oxy compound in the presence of air, which for a nickel(II) complex can be described [233] by Eq. (45) in which, for the sake of simplicity, only the reactant and final product are shown (intermediates involved in the course of the reactions are omitted in the equation).

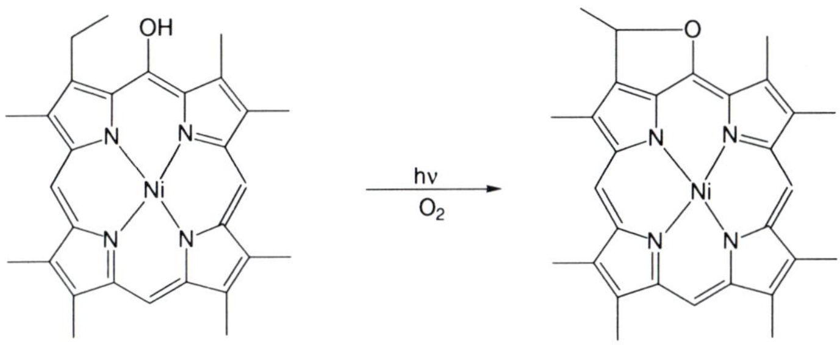

(45)

A similar cyclic ether formation was observed in systems containing octaethyl-α-hydroxyporphyrinato zinc(II) or Ni(II) complexes [244].

Photochemical changes in ring composition are not simple to follow. Due to the possibilities offered by NMR, diamagnetic compounds are predominantly investigated. A number of the fundamental questions of photobiology cannot be answered without a deep and systematic research in this field. One can be sure that, poetically speaking, the lifetime of the research interest in this area will be counted in decades at least.

8 Photochemical Formation and Decomposition of Polynuclear Complexes

The subject of this Section is photochemical processes in which polynuclear tetrapyrrole complexes with metal–metal or metal–bridge–metal bonds are formed from or decomposed to mononuclear tetrapyrrole ligand containing complexes. All known cases belonging to this category are transition metal tetrapyrroles.

The reason of why such reactions are discussed separately stems from the fact that metal–metal (both a direct or a bridge mediated) interaction causes certain specific properties of polynuclear complexes, and cannot be, therefore taken as a simple sum of their independent mononuclear constituents. A particular case is some phthalocyanine dimers in which the monomers are bound via a close π-π system contact of the phthalocyanine ligands. Such dimers are included in this Section too in spite of the fact that there is neither a direct metal–metal bond nor a bridge connection.

Photoinduced formation of binuclear tetrapyrroles (in all instances polynuclear means binuclear in this Section) is a rare process. In reality, it is an addition reaction involving two ground-state species, at least one of them being a coordinatively unsaturated intermediate formed in the primary photochemical step. A true photochemical binuclear complex formation, involving an electronically excited complex with enhanced Lewis acidity, has not been observed so far (the formation of excimers – Sect. 2 – does not fall in the category of polynuclear complex formation as excimers are excited-state species).

It is meaningless trying to find a relation between the nature of excited states and the ability of complexes to undergo photodimerization, since in the corresponding addition reactions ground-state molecules participate. The formation of coordinatively unsaturated species should occur from the excited states involving the central atom as one of the central atom – axial ligand bonds must be cleaved.

The data on the formation and decomposition of polynuclear metallotetrapyrroles are compiled in Table 6.

The formation of the unsaturated intermediate $Ru^{II}(OEP)$ from the hydridocomplex $Ru^{III}H(OEP)$ is supposed to occur from a LMCT excited state. As a consequence, the dimeric $[Ru^{II}(OEP)]_2$ with a metal–metal interaction is formed [245]. Irradiation of some systems containing $Fe^{III}(Por)N_3$ leads to μ-nitrido bridged binuclear mixed-valence complexes $[(Por)Fe^{III}{-}N{-}Fe^{IV}(Por)]$ [134, 162]. In both cases photochemistry was used as a conventional preparative route for synthesis of the binuclear complexes.

Comparing with photochemically induced formations of polynuclear complexes, photodecompositions of such compounds are more frequent processes [1] and this general phenomenon is fully applicable in the photochemistry of metallotetrapyrroles as well.

Three types of metallotetrapyrroles discussed in this Section can undergo polynuclear-to-mononuclear photochemical decomposition:

1) complexes with a direct metal–metal bond;
2) complexes with a metal–bridge–metal bond;
3) complexes in which the bonding of mononuclear entities is mediated by the tetrapyrrole π-systems.

Unlike typical inorganic complexes with simple ligands [1, 98] no information is available on the photoinduced metal–metal bond splitting in polynuclear

Table 6. Photochemical formation and decomposition of polynuclear tetrapyrrole complexes

Reactants	Products	Conditions, parameters of reactions, notes	Ref.
Ru(OEP)H	$[Ru(OEP)]_2$	Solvent: λ_{irr} = visible light; degassed solutions; t = 25°C	[245]
$Rh(OEP)(CH_3)$	$[Rh(OEP)]_2$	Solvent: benzene; λ_{irr} = 532 or 355 nm; $\phi \sim 10^{-1}$–10^{-2} and decreases in the presence of O_2; the singlet S_1 state is responsible for the photoreaction (homolytic Rh–C bond splitting)	[153]
$(\mu\text{-TPP})[Rh(CO)_2]_2$	$[Rh(TPP)]_2$	Solvent: benzene; λ_{irr} = ultraviolet + visible radiation; $\phi = 0$ at $\lambda_{irr} > 420$ nm, and 0.057 at $\lambda_{irr} = 300$ nm; the product is a diamagnetic dimer; higher singlets are photoreactive;	[246]
	Rh(TPP)(CO)Cl	Solvent: benzene + CCl_4; λ_{irr} = unfiltered radiation; $\phi = 0.075$ at $\lambda_{irr} < 320$ nm, and $\sim 10^{-4}$ at $\lambda_{irr} > 420$ nm;	[247]
	Rh(TPP)(py)	Solvent: benzene + pyridine; λ_{irr} = unfiltered radiation; $\phi = 0$ at $\lambda_{irr} > 420$ nm, and 0.055 at $\lambda_{irr} = 337.2$ nm	[248]
$Fe^{III}(Por)N_3$ Por = OEP, TTP	(Por)Fe^{III}–N–Fe^{IV}(Por)	Solid film irradiated by high laser power at $\lambda_{irr} = 413.1$ nm, or THF solution by a high-pressure Hg-lamp (in pyridine $Fe(Por)(py)_2$ is formed); μ-nitrido complex formed from the precursor Fe^{V}(Por)N	[134, 162]
$[Ni(TSPc)]^{4-}$	$[Ni^{II}Ni^{III}(TSPc)(TSPc^{\bullet})]^{6-}$	Solvent: water, pH = 1; the primary radical $[Ni^{II}(TSPc^{\bullet})]^{3-}$ formed also by radiolysis or oxidation with Ce^{4+}	[212]
$Ag_2(TPP)$	Ag(TPP)	Solvent: pyridine; no details given	[246]
(TPP)Fe–O–Fe(TPP)	Fe(TPP) + FeO(TPP)	Solvent: benzene, pyridine, CH_2Cl_2, C_6H_{12}; λ_{irr} = visible light; $\phi \sim 10^{-4}$; Fe^{IV}O(TPP) behaves as a strong oxidant and may have a character of [Fe^{III}O(TPP)] in which the π-cation radical is strongly antiferromagnetically coupled to Fe(III) [250]	[249–253]
(TPPC)Fe–O–Fe(TPPC)	Fe(TPPC) + FeO(TPPC)	Solvent: water; λ_{irr} = visible light; $\phi \sim 10^{-5}$; $Fe^{IV}O(TPPC^{\bullet})$ behaves as an 2-electron oxidant towards amines, or acts as an oxygen atom donor	[254]
$[(TPPS)Fe\text{–}O\text{–}Fe(TPPS)]^{8-}$	2 $[Fe^{II}(TPPS)]^{4-}$	Solvent: water + ethanol; λ_{irr} = unspecified; ethanol oxidized to acetaldehyde	[255]
(TdPc)Fe–O–Fe(TdPc)	2 Fe^{II}(TdPc)	Solvent: DMF or THF; λ_{irr} = white visible light; $\phi \sim 10^{-2}$; products of oxidation unidentified	[256]
(Pc)Mn–O–Mn(Pc)	2 $[Mn^{II}(Pc)(OH)_2]^{2-}$	Solvent: water + ethanol; pH ~ 12; λ_{irr} = white light; $\phi \sim 10^{-2}$	[257]

Table 6. (continued)

Reactants	Products	Conditions, parameters of reactions, notes	Ref.
$[(TSPc)Mn{-}O{-}Mn(TSPc)(OH)]^{9-}$	$[Mn^{II}(TSPc)(H_2O)]^{4-}$ + $[Mn^{III}(TSPc)(H_2O)(O^{\bullet})]^{4-}$	Solvent: water, pH = 11; λ_{irr} = 229 or 254 nm; $[Mn^{III}(TSPc)(H_2O)(O^{\bullet})]^{4-}$ converts to $[Mn^{III}(TSPc^{\bullet})(OH)(H_2O)]^{3-}$ and gives the parent Mn^{III} dimer by a reverse reaction; Mn^{IV} oxo complex formation not considered	[130]
$[(TTP)Nb^{V}(\mu{-}O)_3Nb^{V}(TTP)]$	$[Nb^{V}O(TTP)(O_2^{\bullet})]$ + $[Nb^{V}O(TTP)(O^{\bullet})]$	Solvent: benzene; the presence of O_2; λ_{irr} = visible light; the primary photochemical step is a homolytic cleavage of one of the Nb–O bonds; the radical products are stable for some days in the absence of scavengers	[257]
$[Fe(TSPc)(H_2O)_2]_2^{8-}$	$[Fe^{II}(TSPc^{\bullet})]^{5-}$ + $[Fe^{II}(TSPc^{\bullet})]^{3-}$	Solvent: water, pH ~ 1–3; λ_{irr} > 220 nm; $\phi \sim 10^{-3}$–10^{-5}; $[Fe^{II}(TSPc^{\bullet})]^{5-}$ converts to $Fe^{I}(TSPc)^{5-}$	[150]
$[Co^{III}(TSPc)(H_2O)_2]_2^{6-}$	$[Co^{III}(TSPc^{\bullet})(H_2O)_2]^{2-}$ + $[Co^{III}(TSPc^{\bullet})(H_2O)_2]^{4-}$	Solvent: water, pH ~ 1–3; λ_{irr} > 220 nm; $[Co^{III}(TSPc^{\bullet})(H_2O)_2]^{2-}$ reacts with 2-propanol producing $[Co^{II}(TSPc)(H_2O)_2]^{4-}$	[150]
$[Co^{II}(TSPc)]_2^{8-}$	$[Co^{I}(TSPc)]^{5-}$ + $[Co^{III}(TSPc)]^{3-}$	Solvent: water, pH ~ 1–4; λ_{irr} = 225–550 nm; $\phi \sim 10^{-3}$; $Co^{I}(TSPc)^{5-}$ reacts with O_2 forming H_2O_2	[258]
$[Cu^{II}(TSPc)]_2^{8-}$	$Cu^{II}(TSPc^{\bullet})]^{5-}$ + $[Cu^{II}(TSPc^{\bullet})]^{3-}$	Solvent: water, pH ~ 1–4; λ_{irr} = 225–550 nm; $\phi \sim 10^{-3}$; $[Cu^{II}(TSPc^{\bullet})]^{3-}$ is reduced by 2-propanol; $[Cu^{II}(TSPc^{\bullet})]^{5-}$ with O_2 forms $[Cu^{II}(TSPc{-}O_2^{\bullet})]^{5-}$	[258]
$[Cu^{II}(OSPc)]_2$	$[Cu^{I}(OSPc)]^{-}$ + $[Cu^{III}(OSPc)]^{+}$	Solvent: $CHCl_3$; λ_{irr} > 280 nm; $Cu^{I}(OSPc)^{-}$ is oxidized to $Cu^{III}(OSPc)^{+}$ by $CHCl_3$	[213, 259]
1,5-di-[Coβ – cob]-pentane	Co^{I}-cobalamin + Co^{II}-cobalamin	Solvent: D_2O; λ_{irr} = sunlight; the dinuclear reactant understood as a latent pentyl-1,5-diradical	[122]

metallotetrapyrroles (class 1 of the above classification). There is no reason, however, why such reactions should not be discovered in the future.

The second class of compounds comprises complexes in which the central atoms are connected through μ-oxo bridge(s) or the tetrapyrrole ligand itself acts as a bridge. The latter of the two groups is represented by the complex (μ-TPP) $[Rh^{I}(CO)_2]_2$ [246–248]. The complex is photochemically reactive only when being excited into higher singlet states (S_2, S_3, etc.) as deduced of the dependence of the quantum yield values on the wavelength of incident light. The decomposition of the final products and the oxidation number of the central rhodium atom in them are determined by the character of solvent used.

In noncoordinating solvents (e.g. benzene) as a final product possessing TPP-ligands, the dimer $[Rh^{II}(TPP)]_2$ was found. In the presence of redox stable coordinating solvents (e.g. pyridine) the monomeric pentacoordinated Rh^{II}(TPP) (py) was formed. In the presence of redox reactive solvents (e.g. CCl_4) the complex Rh^{III}(TPP)(CO)Cl was produced. The primary photochemical step is probably identical in all the systems. Its stoichiometry can be expressed as follows

$$(\mu\text{-TPP})[Rh^{I}(CO)_2]_2 \xrightarrow{h\nu} Rh^{II}(TPP) + Rh^{0} + 4CO \qquad (46)$$

The formation of Rh^0 is a consequence of the homolysis of a Rh–N (tetrapyrrole) bond. In a similar way the disproportionation of excited Ag^{I}_2(TPP) to Ag^{II}(TPP) and Ag^0 presumably occurs [246].

Excited states responsible for the photodisproportionation of μ-oxo-bridged dinuclear complexes (Por)M^{III}–O–M^{III}(Por), where M = Fe or Mn, are believed to be of LMCT nature. The transfer of an electron O2p → M3d leads to the reduction of this central atom and decomposition of the binuclear entity. The second central atom as a part of an oxo compound will be in a higher oxidation state, the whole process is then a disproportionation. In this sense the photochemical decomposition of the complex (TPP)Fe^{III} (μ-O^{-II})Fe^{III}(TPP) is illustrative [249–253].

$$(TPP)Fe\text{–}O\text{–}Fe(TPP) \xrightarrow[LMCT]{h\nu} Fe^{II}(TPP) + Fe^{IV}O(TPP) \qquad (47)$$

The reaction is endergonic and its primary products can recombine to the parent binuclear complex with high efficiency. Such a reverse process is indeed realized if other possible reaction partners are missing in irradiated systems. In the presence of reducing agents, the mononuclear oxometal complex FeO(TPP) behaves as a strong oxidant via the oxygen atom or electron transfer mechanisms. A classical example of the oxygen transfer process is the reaction of FeO(TPP) with PPh_3 yielding Fe^{II}(TPP) and $OPPh_3$ (the reaction itself is a spontaneous ground state process involving photochemically created FeO(TPP)). In this way steady-state photolysis of μ-oxo complexes in the presence of reducing (oxidizing) agents can lead to the formation of only one final tetrapyrrole complex.

The third class of dimers undergoing a photodecomposition to monomers are metallophthalocyanines. Many of them, even at low concentration, associate to dimers. Based on spectral data, associative equilibrium constants were estimated [4] to be of the order of 10^4–10^7. The forces which hold mononuclear entities together are still not fully understood. Usually it is claimed that there is a strong interaction between the π electronic systems. Such a strong interaction of the charge-transfer nature seems to be a privilege of phthalocyanines. The lack of knowledge on the ground-state electronic structure of dimers reflects itself in the description of the electronic structure of their photoreactive excited states. For a given dimer the photoreactive excited state and the course of the photodecomposition are determined by the nature of the metal center (compare, e.g. Co(II) and Co(III) dimers with the same phthalocyanine ligand [150, 258]) and by the ring substituents (see. e.g. the differences in the behaviour of $[Cu^{II}(OSPc)]_2$ and $[Cu^{II}(TSPc)]_2^{8-}$ [213, 258, 259]). The observed differences may be a consequence of the different ability of the phthalocyanine ligands to bear the positive (in π-cation radicals) or negative (in π-anion radicals) charge. In addition to the influence of the ring substituents and the central atom, the stability of individual states can be delicately balanced by the axial coordination. The photoredox decompositions of metallophthalocyanine dimers can be summarized (for details see Table 6) as follows

$$[M^n(Pc)]_2^{2a-} \xrightarrow{h\nu} \begin{cases} [M^{n+1}(Pc)]^{1-a} + [M^{n-1}(Pc)]^{(1+a)-} \\ [M^n(Pc^{\bullet})]^{-1a} + [M^n(Pc^{\bullet})]^{(1+a)-} \end{cases} \tag{48}$$

Thus, a redox photodisproportionation occurs either at the metal atoms or the phthalocyanine rings.

Both kinds of monomeric products can be distinguished if the energy differences in the pairs involved are high enough. If it is low, an interconversion of both forms of oxidized and reduced products can occur. For the first time such a conversion was documented on the electrochemically generated nickel tetraphenylporphyrin possessing a "moving hole" [260]

$$Ni^{II}(TPP) \xrightarrow{-e} [Ni^{II}(TPP^{\bullet})]^+ \rightleftarrows [Ni^{III}(TPP)]^+ \tag{49}$$

Such different forms when long-lived are distinguishable by spectral and EPR techniques.

The results on the studies of the formation and decomposition of polynuclear complexes have a great theoretical importance. In addition, knowledge obtained in this field can be applied in practical chemistry too, e.g. in optimization of photocatalytic processes.

9 Applications of Photochemistry of Metallotetrapyrroles

Along with their biological and theoretical importance, metallotetrapyrroles are compounds with a wide spectrum of applications. They are used as catalysts for electrochemical processes, preparation of organic compounds, destruction of water and air pollutants, polymerization, and isomerization processes. They function as pigments, electrochromic materials, constituents of holographic systems, analytical and therapeutical reagents.

The most important natural photochemical process is undoubtedly photosynthesis. Exploitation of artificial photophysical and photochemical processes of tetrapyrrole metal complexes has been focused on the following areas:

1) transformation of organic compounds (oxidations, reductions, isomerizations, decompositions of diazonium compounds) and pollutants (NO_3, NO_2) catalyzed by electronically excited metallotetrapyrroles;
2) conversion of solar energy to chemical and electrical energies;
3) photodiagnosis in medicine and photodynamic therapy;
4) electrophotographic photoreceptors and copying systems;
5) photodynamic damage of plant tissues.

The first of the above classes comprises a large number of phototransformations of organic compounds (Table 7). Of the substrate photoreductions catalyzed by metallotetrapyrroles, the photoreduction of CCl_4 by alcohols has been studied in detail [139, 140, 261]. The reaction can be expressed as follows

$$CCl_4 + CH_3CH(OH)R \xrightarrow[Fe^{III}(Por)]{h\nu} CHCl_3 + CH_3COR + HCl \tag{50}$$

The reaction proceeds by a chain mechanism involving various radical intermediates. Ferric tetraarylporphyrins act as effective light-absorbing catalysts. Using the complex Fe(TDClP)Cl and isopropanol as a solvent, a turnover number, TN, as high as 130 000 was achieved.

Oxidations of hydrocarbons (cycloalkanes, cycloalkenes, aromatics) photocatalyzed by metallotetrapyrroles lead to the formation of epoxides, aldehydes, ketones, alcohols, and carboxylic acids both in solutions and polymer matrices. These processes frequently occur as selective (one-product formation) reactions. Irradiation with visible light has a pronounced accelerating effect on such important industrial processes as the oxidation of thiols to disulfides (Merox process [265]) in a treatment of petroleum distillates or waste water cleaning.

Though the investigation of photocatalytic oxygenations performed of the laboratory scale are often motivated by attempts to understand and mimic the catalytic cycle of cytochrome P450 (a natural catalyst of monooxygenation reactions), the results obtained [159, 253, 266] could be applied to industrial processes as well.

Reactions photocatalyzed by metallotetrapyrroles can be of both a carbon–carbon bond splitting and bond forming nature. The former case applies to many substituted 1,2-diols [262, 263], producing aldehydes, ketones, and carboxylic acids when irradiated in the presence of ferric tetrapyrroles. The latter case can be exemplified by a polymerization of alkyl methacrylates [115] in the presence of $Al(TPP)CH_3$. In both instances the reactions are specific and can be controlled to occur with yields exceeding 90%.

Table 7. Photochemical transformations catalyzed by meatllotetrapyrroles as light absorbing species

Catalyst	Reactants	Products	Turnover number TN, yields, notes	Ref.
Fe(Por)Cl Por = OEP, TDClP, TMsP, TFPP, TPP	$CCl_4 + CH_3CH(OH)R$ R = H or alkyl	$CHCl_3 + CH_3COR$	TN ~ 10^2–10^5 (130 000 for TDClP); $\phi \sim 10^{-2}$; $\lambda_{irr} = 350$–450 nm; solvent: $CH_3CH(OH)R$	[139, 140, 261]
Fe(Por)Cl Por = TTP, TDClP	Cyclohexene + O_2	2-cyclohexene-1-ol 2-cyclohexene-1-one	Iron porphyrins entrapped in a polystyrene matrix; $\lambda_{irr} > 350$ nm; TN ~ 10^3; $\phi \sim 1$	[141]
$Fe(Por)Cl_5$ Por = TClMP, TMPyP	$R_2C(OH)$–$C(OH)R' + O_2$ R = R′ = H, CH_3, C_6H_5	RCHO, RCOOH	Solvent: ACN + H_2O, or H_2O; $\lambda_{irr} > 400$ nm; cofacially hindered, robust $Fe(TClMP)Cl_5$ is more substrate specific than $Fe(TMPyP)Cl_5$	[262]
Fe(Por)X X = Cl^-, ClO_4^- Por = TPP, TMP	Arylsubstituted diols + O_2	Arylaldehydes, arylalkanes	Solvent: DMSO, DMF, pyridine, CH_2Cl_2; $\lambda_{irr} > 420$ nm; 19 diols studied; the products are selectively formed in more than 90% yield in 10 cases.	[263]
$[Fe(TPP)]_2O$	Cycloalkanes + O_2	Epoxides, alcohols, acids, ketones	Solvent: benzene; $\lambda_{irr} = 350$–440 nm; TN ~ 10^1–10^3 (4233 for 1,5-dimethylcycloocta-1,5-diene); selective conversion of cyclooctane, 1-methylcyclooctane, and 1,5-dimethylcycloocta-1,5-diene to the corresponding epoxides	[253]
$Mn(TPP)NO_2$	$PPh_3 + NO_2$	$OPPh_3$ + NO	Solvent: benzene; $\lambda_{irr} = 350$–420 nm; $\phi \sim 10^{-4}$; systems promising for NO_3^- and NO_2^- reduction	[158]
$Mn(TPP)IO_4$	O_2 + toluene, cyclohexene or cyclopentane	Benzaldehyde, cyclohexenone, epoxide, or cyclopentanone	Solvent: benzene; $\lambda_{irr} = 310$–490 nm; selective conversion of toluene to benzaldehyde, and cyclopentane to cyclopentanone; using ClO_4^- the reactions are stoichiometric	[159]
$Sn(TPP)L_2$ L = OH^-, H_2O, $OPPh_3$	$PPh_3 + H_2O + MV^{2+}$	$OPPh_3 + H^+ + MV^+$	Solvent: ACN + H_2O; $\lambda_{irr} = 420$ nm; $\phi \doteq 0$ for L = PPh_3, and 0.3 for L = OH^-; the triplet Sn(II) complexes quenched oxidatively by MV^{2+}	[264]
$Sn(TPP)Cl_2$	$(CH_3)_2N$–C_6H_4–N_2^+	$(CH_3)_2N$–C_6H_5	Solvent: ACN + $C_2H_4Cl_2$; $\lambda_{irr} = 560$–600 nm; $\phi = 1.58$; an exciplex involved in the decomposition of diazonium salts	[48]
$Al(TPP)CH_3$	Methyl methacrylate (monomer)	Polymethyl methacrylate	Solvent: benzene, CH_2Cl_2; $\lambda_{irr} > 420$ nm; 100% conversion can be achieved; in the presence of butyl methacrylate a copolymer is formed	[115]
Zn(PcTSM)	R–C_6H_4–N_2^+ R = CH_3O, $(C_2H_5)_2N$, CH_3	R–C_6H_5	Solvent: ACN; $\lambda_{irr} = 671$ nm; $\phi \sim 10^{-1}$–10^{-2}; triplet exciplexes formed	[54]

It seems to be obvious at present that the spectral properties, stability, and possibility to tune their properties on purpose make metallotetrapyrroles hot candidates for a broader transfer of the theoretical knowledge into industrial photocatalytic processes as well as into the field of the protection of the environment.

The gradual exhaustion of the world resources of fossile fuels and the increase in their prices in the early 1970s prompted world-wide research to develop alternative sources of energy.

The exploitation of solar energy is an especially attractive field as this resource is very rich, infinitely abundant (2×10^{24} J/year), and environmentally clean. Solar energy can be converted into thermal, electrical, and chemical energies. The most ideal system would be that in which water is split into hydrogen and oxygen (an energy rich and ecologically the purest fuel mixture) with a high efficiency under the action of solar radiation due to the photocatalytic effect of an ideal photocatalyst. In this relation two classes of metal complexes are being intensively investigated, namely metallotetrapyrroles, satisfactorily playing their role in photosynthesis, and polypyridine complexes of some transition metals [22].

At the first stage of the research, several zinc(II) tetrapyrroles (porphyrin, phthalocyanine and naphthalocyanine derivatives) were investigated [27] in the systems containing sacrificed compounds. The results obtained contributed to the development of the theory of redox reactions and energy-transfer processes, they were, however, of little practical significance.

Lately the research has been directed to long-range electron-transfer processes in systems possessing metallotetrapyrroles as electron donors (various so-called dyads, triads and polyads have been synthetized and investigated [6]), and to heterogeneous systems, as well as systems combining both approaches (charge separation on the phase boundary or membranes, systems with electrode-like metal or metal oxide particles, micellar systems, etc. [188–191]). Zinc and magnesium compounds belong still to the ones most intensively studied.

The research done up to now contributed significantly to the knowledge on mechanism of photosynthesis, to the theory of the phase transitions, charge and energy transfer, etc. The systems are, unfortunately, far from well-functioning, applicable systems converting solar energy to chemical energy.

Photoelectrochemical and photovoltaic cells are devices converting radiation energy to electrical energy. Tetrapyrrole complexes, usually deposited on electrodes as thin layers, develop photocurrent or photopotential under irradiation. In connection with the theoretical and practical problems of the conversion of solar to electrical energy (the questions of the efficiency of the conversion, spectral sensitivity of metallotetrapyrroles, stability and longevity of the layers, influence of the molecular structure and solid-phase packing on the photoconductive gain of the layers, construction and design of the cells, are still a challenge for researchers in this field) various complexes of chlorins [267], porphyrins [216, 268–270], phthalocyanines [265, 270–273] and naphthalocyanines [265] were investigated. On a laboratory scale the results obtained are really promising. The highest photoconductivity gain (defined as a number of charge carriers passing through the sample per one absorbed photon) reached the value of 4×10^{-2} (using polymeric $[Fe(Pc)(tz)]_n$ at 2000 V cm^{-1} [271]), photocurrent approached to 5×10^{-4} A cm^{-2} (using the liquid crystal porphyrin Zn(OOOP) [269]), photopotentials of several hundreds of mV were obtained [265]. The research is in progress and possibilities of improving the practical characteristics of photoelectrochemical and photovoltaic devices are being searched.

The photochemistry and photophysics of metallotetrapyrroles have significant potential applications in medicine. These applications can be grouped into three areas [274] namely photodynamic therapy, photodiagnosis, and photoprotection.

Photodynamic therapy is a therapeutic method based on the dye-sensitized photooxidation of undesirable biological matter in the target tissue [275, 276]. To induce a photodynamic effect, a sensitizer (in an ideal case, selectively retained in a tumor cell), light of the appropriate wavelength (usually delivered from a dye laser via an optical fibre to the required place) and dioxygen must be present simultaneously. In a simplified way, the sequence of the processes in photodynamic therapy can be expressed by three steps: the photoexcitation of a sensitizer; generation of singlet oxygen $^1O_2$ in the energy transfer process between the excited sensitizer and ground-state triplet dioxygen molecules; oxidation of a biological matter in the cell by singlet oxygen.

An excellent sensitizer should be red light absorbing, nontoxic, selectively retained in malignant cells relative to normal adjacent tissue, an efficient generator of cytotoxic ($^1O_2$) species, luminescent for vizualization, and water-soluble. Among the sensitizers studied on the molecular (chemical), cellular (biological) levels and screened on the clinical (medical) level, also metallotetrapyrroles have been used. "In vitro" and "in vivo" investigations have been performed [275–279] with sulphonated phthalocyanines (with Al(III), Ga(III), In(III), Zn(II) as the central atoms), prophyrin derivatives of Zn(II), Ga(III), In(III), naphthalocyanines of Si(IV), etc. At the clinical level, several thousand patients have been treated by photodynamic therapy. The use of metallotetrapyrroles as drugs is still in its infancy. Nevetherless, exciting research and effective applications are expected in the field of photodynamic therapy which should become a promising curative method.

Photodiagnosis is a method based on selective excitation of luminescence emitted by a compound (e.g. Zn(PPIX)) accumulated in blood (e.g. in a treatment of iron-deficiency anemia) or cells (e.g. in a treatment of malignant neoplasms, when the photodiagnosis can precede photodynamic therapy).

Photoprotection is an inhibition of heme photosensitization via the quenching of photoreactive intermediates ($^1O_2$) by orally administrated β-carotene. Photoprotection concerns some kinds of porphyrias (the hereditary or chemically induced production of excessive amount of porphyrins in human and animals). Similarly as in photodynamic therapy, a raised concentration of porphyrins (Mb, Hg, cytochromes) causes the destruction of cells in the presence of O_2 which is, however, an undesirable process. β-Carotene as an effective quencher of singlet oxygen protects patients from the disease.

It should be mentioned that the impact of excited metallotetrapyrroles on biological systems is not limited to the above areas. It was found and patented [280] that some water-soluble phthalocyanine complexes exhibit bactericidal action on wet cotton fabric exposed to air and solar radiation. Zn(TPPS) is one of the best sensitizers in photokilling the unarmoured dinoflagellate *Ptychodiscus brevis*. In the future further directions of photobiological applications of metallotetrapyrroles may emerge.

Charge generating and transporting behavior of irradiated metallotetrapyrrole layers is exploited in the function of colour copiers [281–284]. Suited and practically used electrophotographic photoreceptors in the copiers comprise TiPc, TiOPc, VOPc, MnPc as parts of the layers.

The photodynamic effect, besides its use in medicine and bacterial damage processes, can be advantageously applied in agriculture as well. In the last decade a new kind of herbicides, named according to their action, tetrapyrrole-dependent photodynamic herbicides [285] were synthetized, investigated, and applied. These herbicides are polycomponent substances which force green plants to accumulate an overcritical amount of tetrapyrroles (e.g. chlorophyll derivatives and their metabolic intermediates). Due to solar energy absorption the accumulated tetrapyrroles photosensitize the generation of singlet oxygen which consecutively damages the treated plants via killing their cellular membranes by oxidation.

Photodynamic herbicides themselves do not contain metallotetrapyrroles but they are responsible for their formation and, subsequently, for their photodamage action. These herbicides have been successfully applied to more than 10 weed species.

10 Concluding Remarks

Excited-state chemistry of tetrapyrroles and their complexes is a branch of chemistry with a strong linking to other chemical disciplines, biology and physics. This branch draws motivations for the own development from the above areas of science and technology and it itself acts as a rich source of stimulation for these and further areas.

The development of the photochemistry of metallotetrapyrroles is, in general, very rapid. The rate and acceleration of this development for individual groups of complexes, scientific tasks solved, and applicative fields, are not, however, equivalent. It is hardly possible to cover and assess all these aspects in one article. It is nevertheless hoped that this review offers, at least partly, answers to the questions: "what has been done in recent years" and "what can be expected in near future" in the photochemistry of artificial metallotetrapyrroles.

Acknowledgments. I am grateful to Professor K. Kalyanasundaram for giving me a copy of his monograph "Photochemistry of Polypyridine and Porphyrin Complexes" [22] which was issued just at that time when I was completing the manuscript of this review.

11 References

1. Sýkora J, Šima J (1990) Photochemistry of coordination compounds, Elsevier/Veda, Amsterdam/Bratislava
2. Buchler JW (ed) (1987) Struct Bonding 64: 1
3. Buchler JW (ed) (1991) Struct Bonding 75: 1
4. Lezonoff CC, Lever ABP (eds) (1989) Phthalocaynines – properties and applications, VCH, New York
5. Scheer H (ed) (1991) Chlorophylls, CRC Press, Boca Raton
6. Gust D, Moore TA (1991) Top Curr Chem 159: 103
7. Michel-Beyerle ME (ed) (1990) Reaction centers of photosynthetic Bacteria. Springer, Berlin Heidelberg New York
8. Fajer J (1991) Chem Ind 869
9. Tiede DM, Washishta ACJ (1991) Mol Cryst Liq Cryst 194: 191
10. Wasielewski MR, Gaines GI, O'Neil MP, Svec WA, Niemczyk MP (1991) Mol Cryst Liq Cryst 194: 201
11. McLondon G, Hake R, Zhang Q, Corin A (1991) Mol Cryst Liq Cryst 194: 225
12. Qin L, Rodgers KK, Sligar SG (1991) Mol Cryst Liq Cryst 194: 311
13. Zhang Q, Hake R, Billstone V, Simmons J, Falvo J, Hozschu D, Lu K, McLondon G, Corin A (1991) Mol Cryst Liq Cryst 194: 343
14. Hakr R, Zhang Q, Marohn J, McLendon G, Corin A (1991) Mol Cryst Liq Cryst 194: 351
15. Rizzarelli E, Theophanides T (eds) (1991) Chemistry and properties of biomolecular systems, Kluwer, Dordrecht
16. Harriman A, Nowak AK (1990) Pure Appl Chem 62: 1107
17. Lavellee DK (1987) The Chemistry and Biochemistry of N-substituted Porphyrins, VCH, Weinheim
18. Govindjee G (ed) (1987) Excitation energy and electron transfer in photosynthesis, Martinus Nijhoff, Dordrecht
19. Barber J, Malkin R (eds) (1989) Techniques and new development in photosynthesis research, Plenum, New York
20. Mauzerall D (1978) In: Dolphin D (ed) The porphyrins, vol. V, Acadmic Press, New York, p 29
21. Hopf FR, Whitten DG (1978) In: Dolphin D (ed) The porphyrins, vol. II, Academic Press, New York, p 161
22. Kalyanasundaram K (1992) Photochemistry of polypyridine and porphyrin complexes, Academic Press, New York
23. Letokhov VS (ed) (1987) Laser picosecond spectroscopy and photochemistry of biomolecules, Adam Hilger, Bristol
24. Gouterman M, Rentzepis PM, Straub KD (eds) (1986) Porphyrins: Excited States and Dynamics, ACS Symp Ser 321
25. Blauer G, Sund H (1985) Optical properties and structure of tetrapyrroles, deGruyter, Berlin
26. Serpone N, Jamieson MA (1989) Coord Chem Rev 93: 87
27. Darwent JR, Douglas P, Harriman A, Porter G, Richoux MC (1982) Coord Chem Rev 44: 83
28. Sayer P, Gouterman M, Connel CR (1982) Acc Chem Res 15: 73
29. Šima J (1992) Comments Inorg Chem 13: 227
30. Kapinus EI (1988) Photonics of molecular complexes, Dumka Naukova, Kyev (in Russian)
31. Kapinus EI, Aleksankina MM (1990) Zh Fiz Khim 64: 2625
32. Kapinus EI, Aleksankina MM, Starij VP, Lampeka JD, Dilung II (1988) Khim Vys Energ 22: 456
33. Kapinus EI, Starij VP, Dilung II (1982) Teor Eksp Khim 18: 450
34. Angerhofer A, Toporowicz M, Bowman MK, Norris JR, Levanon H (1988) J Phys Chem 92: 7171

35. Roy JK, Whitten DG (1971) J Am Chem Soc 93: 7093
36. Lopp IG, Hendren RW, Wildes PD, Whitten DG (1970) J Am Chem Soc 92: 6440
37. Roy JK, Carroll FA, Whitten DG (1974) J Am Chem Soc 96: 6349
38. Kapinus EI, Aleksankina MM, Starij VP, Dilung II (1989) Dokl Akad Nauk SSSR 278: 1165
39. Kapinus EI, Aleksankina MM, Starij VP, Boghillo VI, Dilung II (1985) J Chem Soc Faraday Trans 81: 631
40. Renge IV, Kuzmin VA, Mirinov AF, Borisevich JE (1982) Dokl Akad Nauk SSSR 263: 143
41. Becker HGO, Lehmann T, Zieba J (1989) J Prakt Chem 331: 806
42. Becker HGO, Lehmann T, Schütz R (1985) J Prakt Chem 327: 21
43. Shakhverdov PA (1971) Opt Spektrosk 30: 81
44. Harriman A, Porter G, Searle N (1979) J Chem Soc Faraday Trans 75: 1515
45. Gore BL, Harriman A, Richoux MC (1982) J Photochem 19: 209
46. Hoshino M, Seki H, Shizuka H (1985) J Phys Chem 89: 470
47. Hoshino M, Hirai T (1987) J Phys Chem 91: 4510
48. Becker HGO, Grossmann K (1990) J Prakt Chem 332: 241
49. Turpin PY, Chinskij L, Laigle A, Tsuboi M, Kincaid JR, Nakamoto K (1990) Photochem Photobiol 51: 519
50. Mercer-Smith JA, Sutcliffe CR, Schmell RM, Whitten DG (1979) J Am Chem Soc 101: 3995
51. Sapunov VV (1988) Khim Fiz 7: 1215
52. Degtiarev LS, Kapinus EI, Skuridin EJ (1984) Khim Vys Energ 18: 56
53. Kapinus EI, Starij VP, Dilung II (1981) Teor Eksp Khim 17: 100
54. Becker HGO, Krüger R, Schütz R (1986) J Prakt Chem 328: 729
55. Becker HGO, Kehlen H, Krüger R (1989) J Prakt Chem 331: 989
56. Prasad DR, Ferraudi G (1983) Inorg Chem 22: 1672
57. Kasuga K, Marimoto H, Ando M (1986) Inorg Chem 25: 2478
58. Andreeva NE, Chibisov AK (1979) Teor Eksp Khim 15: 668
59. Frink ME, Ferraudi G (1986) Chem Phys Lett 124: 576
60. Frink ME, Griger DK, Ferraudi G (1986) J Phys Chem 90: 1924
61. Callis JB, Knowles JM, Gouterman M (1973) J Phys Chem 77: 154
62. Sapunov VV, Tsvirko MP (1980) Opt Spektrosk 49: 283
63. Sapunov VV, Tsvirko MP (1976) Dokl Akad Nauk Byeloruss SSR 20: 208
64. Tait CD, Holten D, Barley MH, Dolphin D, James BR (1985) J Am Chem Soc 107: 1930
65. Kim D, Holten D (1983) Chem Phys Lett 98: 584
66. Kim D, Kirmaier Ch, Holten D (1983) Chem Phys 75: 305
67. Yamaji M (1991) Inorg Chem 30: 2949
68. Yamaji M, Hama Y, Hoshino M (1990) Chem Phys Lett 165: 309
69. Hoshino M, Kogure M (1989) J Phys Chem 93: 5478
70. Dixon DW, Kirmaier Ch, Holten D (1985) J Am Chem Soc 107: 808
71. Lavalette D, Tetreau C, Momenteau M (1979) J Am Chem Soc 101: 5395
72. Momenteau M, Lavellette D (1978) J Am Chem Soc 100: 4322
73. Tait CD, Holten D, Gouterman M (1984) J Am Chem Soc 106: 6653
74. Hoshino M, Kogure M, Amano K, Hinohara T (1989) J Phys Chem 93: 6655
75. Kim D, Spiro TG (1986) J Am Chem Soc 108: 2099
76. Findsen EW, Alston K, Shelnutt JA, Ondrias MR (1986) J Am Chem Soc 108: 4009
77. Shelnutt JA, Alston K, Findsen EW, Ondrias MR, Rifkind JM (1986) ACS Symp Ser 321: 232
78. Farrell N, Dolphin DH, James BR (1978) J Am Chem Soc 100: 324
79. Paulson DR, Bhakta SB, Hyun RY, Yuen M, Beaird CE, Lee SC, Kim I, Ybarra J (1983) Inorg Chim Acta 151: 149
80. Vogler A, Kunkely H (1976) Ber Bunsenges 80: 425
81. Hoshino M, Kashiwagi Y (1990) J Phys Chem 94: 673
82. Whitten DG, Eaker DW, Horsey BE, Schmehl RH, Worsham PR (1978) Ber Bunsenges Phys Chem 82: 858
83. Hopf FR, Whitten DG (1976) J Am Chem Soc 98: 7422
84. Crawford BA, Ondrias MR (1989) J Phys Chem 93: 5055
85. Sovocool GW, Hopf FR, Whitten DG (1972) J Am Chem Soc 94: 4350
86. Hopf FR, O'Brien TP, Scheidt WR, Whitten DG (1975) J Am Chem Soc 97: 277
87. Antipas A, Buchler JW, Gouterman M, Smith PD (1978) J Am Chem Soc 100: 3015
88. Rillema DP, Nagle JK, Barringer LF, Meyer TJ (1981) J Am Chem Soc 103: 56
89. Collman JP, Barnes CE, Collins TJ, Brothers PJ (1981) J Am Chem Soc 103: 7030
90. Chow BC, Cohen IA (1971) Bioinorg Chem 1: 57

91. Hoshino M, Yasufuku K (1985) Chem Phys Lett 117: 259
92. Vogler A, Kisslinger J, Buchler JW (1985) in Blauer G, Sund H (eds) Optical properties and structure of tetrapyrroles, 107, deGruyter, Berlin
93. Collman JP, Barnes CE, Woo LK (1983) Proc Natl Acad Sci USA 50: 7684
94. Irwin C, Stynes DV (1978) Inorg Chem 17: 2682
95. Muralidharan S, Ferraudi G, Schmatz K (1982) Inorg Chem 21: 2961
96. Vogler A, Hirschmann R, Otto H, Kunkely H (1976) Ber Bunsenges 80: 420
97. Hirai Y, Murayama H, Aida T, Inoue S (1988) J Am Chem Soc 110: 7387
98. Balzani V, Carassiti V (1970) Photochemistry of coordination compounds, Academic Press, New York
99. Geoffroy GL, Wrighton MS (1979) Organometallic Photochemistry, Academic Press, New York
100. Adamson AW (1967) J Phys Chem 71: 798
101. Milder JS, Bjorling SC, Kuntz ID, Linger DS (1988) Biophys J 53: 659
102. Caldwell K, Noe LJ, Coccone JD, Taylor TG (1986) J Am Chem Soc 108: 6150
103. Gulbinas V, Dzhagarov BM, Kabelka V, Savickene Z (1987) Dokl Akad Nauk SSSR 293: 987
104. Huang Y, Marden MD, Lambry JC, Fontaine-Aupart MP, Pansu R, Martin JL, Poyart C (1991) J Am Chem Soc 113: 9141
105. Lavalette D, Tetreau C, Momenteau M, Mispelter J, Lhoste JM (1990) Laser Chem 10: 297
106. Anfinrud PA, Han C, Hochstrasser RM (1989) Proc Natl Acad Sci USA 86: 8387
107. Hill BC, Marmor S (1991) Biochem J 279: 355
108. Dyer RB, Peterson KA, Stoutland PO, Woodruff WH (1991) J Am Chem Soc 113: 6276
109. Sassaroli M, Ching YC, Argade PV, Rousseau DL (1988) Biochemistry 27: 2496
110. van Eldik R (1992) In: Williams AF, Floriani C, Merbach AE (eds) Perspectives in coordination chemistry, Verlag Helvetica Chimica Acta, Basel, 55
111. Alelyunas YW, Flemming PE, Finke RG, Pagano TG, Marzilli LG (1991) J Am Chem Soc 113: 3781
112. Inoue S, Takeda N (1977) Bull Chem Soc Japan 50: 984
113. Cocolios P, Guilard R, Bayeul R, Lecomte C (1985) Inorg Chem 24: 2058
114. Tero-Kubota S, Hoshino M, Kato M, Goedken VL, Ito T (1985) J Chem Soc Chem Commun 959
115. Kuroki M, Aida T, Inoue S (1987) J Am Chem Soc 109: 4737
116. Murayama H, Inoue S (1985) Chem Lett 1377
117. Kräutler B (1984) Helv Chim Acta 67: 1053
118. Slanina Z (1986) Contemporary Theory of Chemical Isomerization, Academia, Prague
119. Gažo J, Boča R, Jóna E, Kabešová M, Macášková Ľ, Šima J, Pelikán P, Valach F (1982) Coord Chem Rev 43: 87
120. Freitag RA, Mercer-Smith JA, Whitten DG (1981) J Am Chem Soc 103: 1226
121. König E (1991) Struct Bonding 76: 51
122. Kräutler B (1991) Coord Chem Rev 111: 215
123. Martin BD, Finke RG (1992) J Am Chem Soc 114: 585
124. Jin T, Suzuki T, Imamura T, Fujimoto M (1987) Inorg Chem 26: 1280
125. Suslick KS, Watson RA, Wilson RS (1991) Inorg Chem 30: 2311
126. Imamura T, Jin T, Suzuki T, Fujimoto M (1985) Chem Lett 847
127. Hendrickson DN, Kinnaird MG, Suslick KS (1987) J Am Chem Soc 109: 1243
128. Engelsma G, Yamamoto A, Markham E, Calvin M (1962) J Phys Chem 66: 2517
129. Takahashi K, Komura T, Imanaga H (1983) Bull Chem Soc Japan 56: 3203
130. Ferraudi G, Granifo J (1985) J Phys Chem 89: 1206
131. Suslick KS, Bautista JF, Watson RA (1991) J Am Chem Soc 113: 6111
132. Suslick KS, Acholla FA, Watson RA (1989) Int Symp Photochem Synth Catal, Univ Ferrara, 53
133. Rehorek D, Berthold T, Henning H, Kemp TJ (1988) Z Chem 28: 72
134. Buchler JW, Dreher Ch (1983) Z Naturforsch 39b: 222
135. Ogura T, Fidler V, Ozaki Y, Kitagawa T (1990) Chem Phys Lett 169: 457
136. Fidler V, Ogura T, Dato SI, Aoyagi K, Kitagawa T (1991) Bull Chem Soc Japan 64: 2315
137. Ozaki Y, Iriyama K, Ogoshi K, Kitagawa T (1987) J Am Chem Soc 109: 5583
138. Tohara A, Sato M (1989) Chem Lett 153
139. Bartocci C, Maldotti A, Varani G, Battistoni P, Carassiti V, Mansuy D (1991 Inorg Chem 30: 1255
140. Maldotti A, Bartocci C, Amadelli R, Carassiti V (1989) J Chem Soc Dalton Trans 1197

141. Polo E, Amadelli R, Carassiti V, Maldotti A (1992) Inorg Chim Acta 192: 1
142. Bartocci C, Scandola F, Ferri A, Carassiti C (1980) J Amer Chem Soc 102: 7067
143. Maldotti A, Bartocci C, Chiorboli C, Ferri A, Carassiti V (1985) J Chem Soc Chem Commun 881
144. Maldotti A, Bartocci C, Amadelli R, Carassiti V (1983) Inorg Chim Acta 74: 275
145. Bartocci C, Maldotti A, Traverso O, Bignozzi CA, Carassiti V (1983) Polyhedron 2: 97
146. Bartocci C, Maldotti A, Carassiti V, Traverso O, Ferri A (1985) Inorg Chim Acta 107: 5
147. Šima J, Ducárová T, Havašová K, Antalík M (1990) Inorg Chim Acta 176: 15
148. Hoshino M (1985) Chem Phys Lett 120: 50
149. Zakrzewski J, Giannotti C (1991) Photochem Photobiol A 57: 479
150. Ferraudi G (1979) Inorg Chem 18: 1005
151. Hatano K, Usui K, Ishida Y (1981) Bull Chem Soc Japan 54: 413
152. Hoshino M, Imimura Y, Konishi S (1992) J Phys Chem 96: 179
153. Hoshino M, Yasufuku K, Seki H, Yamazaki H (1985) J Phys Chem 89: 3080
154. Yamaji M, Hama Y, Miyazaki Y, Hoshino M (1992) Inorg Chem 31: 932
155. Groves JT, Takahashi T, Butler W (1983) Inorg Chem 22: 884
156. Groves JT, Takahashi T (1983) J Am Chem Soc 105: 2074
157. Boreham CJ, Latour JM, Marchon JC (1980) Inorg Chim Acta 45: L69
158. Suslick KS, Watson RA (1991) Inorg Chem 30: 912
159. Suslick KS, Achlla FA, Cook BR (1987) J Am Chem Soc 109: 2818
160. Imamura T, Yamamoto Y, Suzuki T, Fujimoto M (1987) Chem Lett 2185
161. Yamamoto Y, Imamura T, Suzuki T, Fujimoto M (1988) Chem Lett 261
162. Wagner WD, Nakamoto K (1989) J Am Chem Soc 111: 1590
163. Wagner WD, Nakamoto K (1988) J Am Chem Soc 110: 4044
164. Kraut B, Ferraudi G (1988) Inorg Chim Acta 149: 273
165. Barley M, Dolphin D, James BR, Kirmaier C, Holten D (1984) J Am Chem Soc 106: 3937
166. Barley MH, Becker YJ, Domazetis G, Dolphin D, James BR (1983) Can J Chem 61: 2389
167. Harriman A, Porter G (1980) J Chem Soc Faraday Trans II 76: 1429
168. Povlock SL, Dennis LA, Geiger KD (1990) Inorg Chim Acta 176: 295
169. Nyokong T, Gasyna Z, Stillman (1986) Amer Chem Soc, Symp Ser 321: 309
170. Lerner DA, Ricchiero FR, Gianotti C, Mailard P (1982) J Photochem 18: 193
171. Gasyna Z, Stillman MJ (1990) Inorg Chem 29: 5101
172. Šima J, Ducárová T, Šramko T, Kotočová A (1991) Bull Soc Chem Belges 100: 193
173. Šima J, Grib L, Jalčoviková V (1991) Chem Papers 45: 303
174. Šima J, Ducárová T, Šramko T (1992) Pol J Chem 66: 53
175. Šramko T, Šima J, Fodran P (1992) Acta Chim Hung 129: 215
176. Krijukov AI, Kutschmij SJ (1989) Photochemistry of transition metal complexes, Naukova Dumka, Kyev (in Russian)
177. Baumgartner E, Ronco S, Ferraudi G (1990) Inorg Chem 29: 4747
178. Endicott JF, Kumar K, Schwarz CL, Perkovic MW, Lin WK (1989) J Amer Chem Soc 111: 7411
179. Svastits EW, Dawson JH, Breslow R, Gellman SH (1985) J Am Chem Soc 107: 6427
180. Hoffman BM, Natan MJ, Nocek JM, Wallin SA (1991) Struct Bonding 75: 85
181. Faraggi M, Klapper MH (1988) J Am Chem Soc 110: 5753
182. Kuila D, Natan MJ, Rodgers P, Gingrich DJ, Baxter WW, Arnone A, Hoffman BM (1991) J Am Chem Soc 113: 6520
183. Chang IJ, Gray HB, Winkler JR (1991) J Am Chem Soc 113: 7056
184. McLendon, Guarr T, McGuire M, Simolo K, Strauch S, Taylor K (1985) Coord Chem Rev 64: 113
185. Zhou JS, Rodgers MAJ (1991) J Am Chem Soc 113: 7728
186. Conrad DW, Scott RA (1989) J Am Chem Soc 111: 3461
187. Peterson-Kennedy SE, McGourty JL, Ho PS, Sutoris CJ, Liang N, Zemel H, Blough NV, Margoliash E, Hoffman BM (1985) Coord Chem Rev 64: 125
188. Grätzel M, Kalyanasundaram K (eds) (1991) Kinetics and catalysis in microheterogeneous systems, M. Dekker, New York
189. Serpone N, Pelizzetti E (eds) (1989) Photocatalysis, Wiley, New York
190. Balzani V, Scandola F (1991) Supramolecular Photochemistry, Horwood, Chichester
191. Pelizzetti E, Schiavello M (eds) (1991) Photochemical conversion and storage of solar energy, Kluwer, Dordrecht
192. Serpone N, Jamieson MA, Netzel TL (1981) J Photochem 15: 295

193. Ough E, Gasyna Z, Stillman MJ (1991) Inorg Chem 30: 2301
194. Gasyna Z, Browett WR, Stillman MJ (1984) Inorg Chem 23: 382
195. Bergkamp MA, Chang CK, Netzel TL (1983) J Phys Chem 87: 4441
196. Zaleski JM, Chang CK, Leroi GE, Cukier RI, Nocera DG (1992) J Am Chem Soc 114: 3564
197. Turro C, Chang CK, Leroi GE, Cukier RI, Nocera DG (1992) J Am Chem Soc 114: 4013
198. Seki H (1992) J Chem Soc Faraday Trans 1, 88: 35
199. Berman A, Michaeli A, Feitelson J, Bowman MK, Norris JR, Levanon H, Vogel E, Koch P (1992) J Phys Chem 96: 3041
200. Wöhrle D, Paliuras M (1991) Makromol Chem 192: 819
201. Resch U, Fox MA (1991) J Phys Chem 95: 6169
202. le Roux D, Takakubo M, Mialocq JC (1985) J Chem Phys 82: 739
203. Gerasimov OV, Limar SV, Parmon VN (1991) Photochem Photobiol A 56: 275
204. Hugerat M, Levanon H, Ojadi E, Biczek L, Linschitz H (1991) Chem Phys Lett 181: 400
205. Segawa H, Takehara C, Honda K, Shimidzu T, Asahi T, Mataga N (1992) J Phys Chem 96: 503
206. Brun AM, Harriman A, Heitz V, Sauvage JP (1991) J Am Chem Soc 113: 8657
207. Vergeldt FJ, Koehorst RBM, Schaafsma TJ, Lambry JC, Martin JL, Johnson DG, Wasielewski MR (1991) Chem Phys Lett 182: 107
208. Kadish KM, Maiya GB, Xu QY (1989) Inorg Chem 28: 2518
209. Maiya GB, Barbe JM, Kadish KM (1989) Inorg Chem 28: 2524
210. Satoh M, Ohba Y, Yamauchi S, Iwaizumi M (1992) Inorg Chem 31: 298
211. Tsukahara K, Asami S (1991) Chem Lett 1337
212. Geiger DK, Ferraudi G, Madden K, Granifo J, Rillema DP (1985) J Phys Chem 89: 3890
213. Prasad DR, Ferraudi G (1982) Inorg Chem 21: 2967
214. Nyokong T, Gasyna Z, Stillman MJ (1986) Inorg Chim Acta 112: 11
215. Oliver FW, Thomas C, Hoffman E, Hill D, Sutter TPG, Hambright P, Haye S, Thorpe AN, Quoc N, Harriman A, Neta P, Mosseri S (1991) Inorg Chim Acta 186: 119
216. Kalyanasundaram K, Shelnutt JA, Grätzel M (1988) Inorg Chem 27: 2820
217. Schmatz K, Muralidharan S, Madden K, Fessenden R, Ferraudi G (1982) Inorg Chim Acta 64: L23
218. Bowler BE, Raphael AL, Gray HB (1990) Progr Inorg Chem 38: 259
219. Harriman A, Kubo Y, Sessler JL (1992) J Am Chem Soc 114: 388
220. Batova EE, Levin PP, Shafirovich VJ (1991) Kinet Katal 32: 553
221. Osuka A, Nakajima S, Maruyama K, Magata N, Asahi T (1991) Chem Lett 1003
222. Osuka A, Maruyama K, Magata N, Asahi T, Yamazaki I, Tamai N, Nishimura Y (1991) Chem Phys Lett 181: 413
223. Osuka A, Nagata T, Maruyama K, Magata N, Asahi T, Yamazaki I, Nishimura Y (1991) Chem Phys Lett 185: 88
224. Osuka A, Yamada H, Maruyama K, Magata N, Asahi T, Yamazaki I, Nishimura Y (1991) Chem Phys Lett 181: 419
225. Kräutler B, Stepanek R (1991) Photochem Photobiol 54: 585
226. Kräutler B, Stepanek R (1985) Angew Chem 97: 91
227. Schulthess P, Ammann D, Simon W, Caderas C, Stepanek R, Kräutler B (1984) Helv Chim Acta 67: 1026
228. Kräutler B, Stepanek R, Holze G (1983) Helv Chim Acta 66: 44
229. Kräutler B (1982) Helv Chim Acta 65: 1941
230. Kräutler B, Stepanek R (1983) Helv Chim Acta 66: 1493
231. Iturraspe J, Gossauer A (1991) Photochem Photobiol 54: 43
232. Struck A, Coniel E, Schneider S, Scheer H (1990) Photochem Photobiol 51: 217
233. Fuhrhop JH, Besecke S, Subramanian J, Mengersen C, Riesner D (1975) J Am Chem Soc 97: 7141
234. Fuhrhop JH, Mauzerall D (1971) Photochem Photobiol 13: 453
235. Mosseri S, Mialocq JC, Perly P, Hambright (1991) J Phys Chem 95: 219
236. Richoux MC, Neta P, Christensen PA, Harriman A (1986) J Chem Soc, Faraday Trans 2, 82: 235
237. Krasnovskii AA (1948) Dokl Akad Nauk SSSR 60: 421
238. Krüger W, Fuhrhop JH (1982) Angew Chem 94: 132
239. Whitten DG, Yau JC, Carroll FA (1971) J Am Chem Soc 93: 2291
240. Iakovides P, Simpson DJ, Smith KM (1991) Photochem Photobiol 54: 335
241. Simpson DJ, Smith KM (1988) J Am Chem Soc 110: 2854
242. Smith KM, Simpson DJ, Snow KM (1986) J Am Chem Soc 108: 6834

243. Shimidzu T, Segawa H, Iyoda T, Honda K (1987) J Chem Soc, Faraday Trans 2, 83: 2191
244. Besecke S, Fuhrhop JH (1984) Angew Chem 86: 125
245. Wayland BB, Newman AR (1981) Inorg Chem 20: 3093
246. Hoshino M, Yasufuku K (1985) Inorg Chem 24: 4408
247. Yamamoto S, Hoshino M, Yasufuku K, Imamura M (1984) Inorg Chem 23: 195
248. Huang JW, Ji LN, Su YD, Wang HZ, Hsieh AK (1991) Chinese Chem Lett 2: 279
249. Richman RM, Peterson MW (1982) K Am Chem Soc 104: 5795
250. Peterson MW, Rivers DS, Richman RM (1985) J Am Chem Soc 107: 2907
251. Berthold T, Rehorek D, Henning H (1986) Z Chem 26: 183
252. Guest CR, Straub KD, Hutchinson JA, Rentzepis PM (1988) J Am Chem Soc 110: 5276
253. Weber L, Haufe G, Rehorek D, Henning H (1991) J Chem Soc, Chem Commun 502
254. Peterson MW, Richman RM (1985) Inorg Chem 24: 722
255. Hatano K, Ishida Y (1982) Bull Chem Soc Japan 55: 3333
256. Lever ABP, Licoccia S, Ramaswamy BS (1982) Inorg Chim Acta 64: L87
257. Matsuda Y, Sakamoto S, Takaki T, Muramaki Y (1985) Chem Lett 107
258. Ferraudi G, Srisankar E (1978) Inorg Chem 17: 3164
259. Prasad DR, Ferraudi G (1981) Inorg Chim Acta 54: L231
260. Dolphin D, Niem T, Felton RH, Fujita I (1975) J Am Chem Soc 97: 5288
261. Bartocci C, Maldotti A, Varani G, Carassiti V, Battioni P, Mansuy D (1989) J Chem Soc, Chem Commun 964
262. Ito Y, Kunimoto S, Miyachi S, Kako T (1991) Tetrahedron Lett 32: 4007
263. Okamoto T, Sasaki K, Oka S (1988) J Am Chem Soc 110: 1187
264. Inoue H, Chandrasekaran K, Whitten DG (1985) J Photochem 30: 269
265. Wöhrle D (1991) Chimia 45: 307
266. Wöhrle D, Gitzel J, Krawczyk G, Tsuchida E, Ohno H, Okura I, Nishisaka T (1988) J Macromol Sci, Chem A25: 1277
267. Watanabe T, Machida K, Suzuki H, Kobayashi M, Honda K (1985) Coord Chem Rev 64: 207
268. Wamser CC (1991) Mol Cryst Liq Cryst 194: 65
269. Gregg BA, Fox MA, Bard AJ (1990) J Phys Chem 94: 1586
270. Schlettwein D, Jaeger NI, Wöhrle D (1991) Ber Bunsenges Phys Chem 95: 1526
271. Meier H, Albrecht W, Hanack M (1991) Liq Cryst Mol Cryst 194: 75
272. Butvilas V, Gulbinas V, Urbas A, Vakhnin A (1992) Radiat Phys Chem 39: 165
273. Takeuchi M, Masui M, Mamose Y (1991) Appl Surf Sci 48/49: 517
274. Jori G, Spikes JP (1984) In: Smith KC (ed) Topics in Photomedicine, Plenum Press, New York
275. Henderson BW, Dougherty TJ (1992) Photochem Photobiol 55: 145
276. MacRobert AJ, Phillips D (1992) Chem Ind 17
277. Ji-Xao C, Rong X, Shi-Ming C, Fa-Du L, Kai-Tai C, Huai-Xin C (1991) Cancer Biochem Biophys 12: 103
278. Takemura T, Ohta N, Nakajima S, Sakata I (1991) Photochem Photobiol 54: 683
279. Takemura T, Ohta N, Nakajima S, Sakata I (1991) Photochem Photobiol 54: 137
280. Polany R, Reinert G, Hoelzle G, Pugin A, Vonderwahl R (1979) Ger Offen 2,812,261
281. Yoshioka H, Oda Y, Takadoro H, Fujimaki Y (1991) Jpn Kokai Tokkyo Koho JP 03,37,671 and JP 03,37,672
282. Omori H (1991) Jpn Kokai Tokkyo Koho JP 03,182,763
283. Omori H (1991) Jpn Kokai Tokkyo Koho JP 03,182,764
284. Fujimaki Y, Takadoro H, Yasuhiro O, Hiroshi Y (1989) Jpn Kokai Tokkyo Koho JP 03,37,660
285. Rebeiz CA, Reddy KN, Nandihalli UB, Velu J (1990) Photochem Photobiol 52: 1099

Author Index Volumes 1-84

Printing: Saladruck, Berlin
Binding: Buchbinderei Lüderitz & Bauer, Berlin